MARVELOUS LIGHT

By Grace For Glory Publishing, LLC
Pittsburgh, PA
www.bygraceforglorylit.com

Lord, your judgements are to be feared for your truth does not belong to me
nor to anyone else, but to us all whom you call to share it as a possession.
With terrifying words you warn against regarding it as a private possession,
or we may lose it. Anyone who claims for his own property what you offer
for all to enjoy, and wishes to have exclusive rights to what belongs to
everyone, is driven from the common truth to his own private ideas, that is
from truth to a lie. (Augustine, Confessions 265)

ISBN: 978-0-9987302-1-9

Table of Contents

All visible objects, man, are but pasteboard masks. But in each event – in the living act, the undoubted deed – there, some unknown but still reasoning thing puts forth the mouldings of its features from behind the unreasoning mask. If man will strike, strike through the mask! How can each prisoner reach outside except by thrusting through the wall?

- Herman Melville -
Moby Dick

INTRODUCTION

Once you were a child. Once you knew what inquiry was for. There was a time when you asked questions because you wanted answers, and were glad when you had found them...You have gone far wrong. Thirst was made for water; inquiry for truth. What you now call the free play of inquiry has neither more nor less to do with the ends for which intelligence was given you than masturbation has to do with marriage. (Lewis, The Great Divorce 328)

C.S. Lewis, one of the premier theologians of the twentieth century, wrote the quote above in 1946 as a frustrated critique of the scholastic indeterminacy of his time. He desired that his peers across all disciplines might rid themselves of one particular idea that was very much in vogue at the time: that in spite of, or perhaps due to, humanity's massive strides in accumulating encyclopedic knowledge of our world, we lost the inclination to attempt to make truth statements, to believe that truth was at all knowable.

Philosophically, scientifically, and elsewise, professionals over the past decades feel less and less sure of objective truth. It has become a very slippery concept. We have more anecdotal data today than ever before in history. We have far more access to and interpretation of systematic scientific results and well-reasoned philosophical arguments, yet in conversation amongst the learned, it sometimes seems that far less can be said with conviction. Unfortunately, this loss of objectivity also leads to the general acceptance of subjective truths. In this mode of thought, what is true for me might not be true for you, and I hold onto my own truths obstinately and with a certain degree of self-assured egotism. The loss of objective truth may be leading humanity to unfounded surety in ourselves and our own particular opinions, regardless of how indefensible they may be.

This book does not share its generation's uneasiness concerning objectivity, but instead, it seeks earnestly to say something true about the world in which we live. Humility is essential for the wise, but false humility

1

in the face of overwhelming, comprehensive evidence smacks more of foolishness than wisdom.

I do not pretend to be an expert in any one topic in the following chapters. It is only through the tremendous correlations drawn between the various concepts that I offer anything worth the writing. Other resources, including the references listed at the end of this text, will surely give far greater explanation of the scientific, philosophical, and theological ideas touched upon here. I would encourage the curious reader, after turning the last page of this book, to pursue his or her own research on the concepts that spark curiosity.

Throughout the history of philosophy and theology, the great prophets and thinkers enjoyed and tried their best to explain the experience of instantaneous revelation. In claiming access to unadulterated truth, they have gone on to make some of the boldest statements in history, such as:

> He who studies [this book] will learn that what escaped the Ancients and Moderns God has entrusted to my tongue. On a wondrous day the Holy Spirit blew it into my heart in a single instant, though its writing took many months. (Suhrawardi 162)

> So in the flash of a trembling glance it attained to that which is. At that moment I saw [God's] invisible nature understood through the things which are made. (Augustine, Confessions 127)

The secularist reading these religious citations might be tempted to discard their misguided explanations of and trust in inspiration, but the one who would do so must then contend with the experiences of the greatest theoretical scientists. In a similar fashion, we are told:

> The formation of hypotheses is the most mysterious of all the categories of scientific method. Where they come from, no one knows. A person is sitting somewhere, minding his own business, and suddenly – flash! – he understands something he didn't understand before. (Pirsig 138)

The careful philosopher of science or theology would be quick to recognize, however, "what seems like sudden insight may be misleading, and must be tested soberly when the divine intoxication has passed" (Russell 123-

4). The thinker who would be content to easily accept the insight of unexpected revelation would constantly be blown to and fro by the winds of his own imaginative fancy. Certainly, ideas arrived at by these means without further investigation would hardly be worth recording or sharing, for they are no more than passing thoughts. For this reason, it is encouraging that the two quotes above were tempered with statements from their authors, respectively:

> I did not first arrive at [my revelation] through cogitation; rather, it was acquired through something else. Subsequently I sought proof for it, so that, should I cease contemplating the proof, nothing would make me fall into doubt. (Suhrawardi 2)

> Nevertheless, the one thing that delighted me in Cicero's exhortation was the advice 'not to study one particular sect but to love and seek and pursue and hold fast and strongly embrace wisdom itself, wherever found.' (Augustine, Confessions 39)

Whatever is the mysterious, divine, or personal source of this type of insight, it ought to be universally recognized that no one needs to lend credence to the non-sensical or internally inconsistent. Each revelation must be reflected upon and tested thoroughly in order to establish its value. Only after such review would it be wise to accept the concept, be it scientific, philosophic, or theological.

REALITY'S WINDOW

Students of science, philosophy, and theology are likely to all make the same statement about their chosen field: they would say they are studying reality, peering behind the thick veil of man's current physical constraints to shed light on the deeper and more profound aspects of the world in which we live. As the students of each discipline uttered the words, I would agree with them, each one.

Human reasoning is classed with the most formidable forces in our universe, demonstrating throughout history its power to understand and modify many aspects of nature for humanity's benefit. Each field has its own claim on the current state of man's progress, and it is important for the student of modern knowledge to recognize that fact.

An interesting point might here be made, however. Although science may have to give token acknowledgement to philosophy's validity, it has no requirement to do the same for theology in itself. Without logical necessity for divinity within one's theory of the universe, theology becomes superfluous. Likewise, philosophy must give nod to science's validity within its confines of human understanding, but no such motion toward theology is required. Theology, on the other hand, must recognize philosophy as science does, and of course must give the philosopher's assent to science. The nature of this acceptance will be explored and developed in greater detail in the pages to come, but for now, one thing remains clear: believers in the divine would be foolish to ignore the other ways in which we understand the world.

> I believe that every fact in nature is a revelation of God, is there such as it is because God is such as he is; and I suspect that all its facts impress us so that we learn God unconsciously... From the moment when first we come into contact with the [physical] world, it is to us a revelation of God, his things seen, by which we come to know the things unseen. (MacDonald 227-8)

The perspective expressed above is valuable for the religiously inclined, though perhaps unfortunately uncommon in today's church. In America, the Christian Church is perceived, and rightly so in many cases, to oppose science to the point of rejecting the general benefits of education all together. Because of the use of Darwinian Evolution to explain biological diversity and the Big Bang Theory to explain the origins of the universe (first proposed by Catholic priest Georges Lemaître), religious folks too often lump all science together as one big enemy of God. Extremists of any religion often make the same mistake.

But science as an intellectual pursuit is completely neutral in making judgements. Science, i.e. man's understanding of the physical world, is not intrinsically for or against any worldview. Science is *for* the physically probable and *against* the physically falsifiable, a constantly refined perspective buttressed by innovative studies that constantly rid science of untenable hypotheses and theories. It is as simple as that.

With this in mind, when the religious man looks at the world around him, even in and through the natural laws of science, he should see the glory of the divine. George MacDonald's quote from above continues on to say:

INTRODUCTION

> In its discovered laws, light seems to me to be such because God is
> such. Its so-called laws are the waving of his garments, waving so
> because he is thinking and loving and walking inside them.
> (MacDonald 228)

There it is. The first mention of light: *Marvelous Light.* MacDonald's
quote coincidentally speaks of light and its physical properties, when he could
have referred to any physical phenomenon in the wide world. We see this
'coincidence' time and time again in the relevant literature. And in the above
quote there is now the opportunity to introduce the topic on which the rest
of this book will focus.

Light is phenomenal, in both the scientific and more common use of the
word. That which has allowed us to experience and explore almost all other
aspects of our physical world is itself the primary mystery by which modern
physicists have grown to understand the very basis of everything we
experience daily. Light as wave and light as particle has taught us so much
and revealed far more of what we do not yet know.

Millennia of inconclusive observations have left man wondering what
light is. Because of its ethereal and ephemeral qualities, there has been little
to definitively say about light for much of history, at least physically. Where
physics could not yet speak, theology and philosophy felt inclined to weigh
in. We have a wealth of historical texts that tell us much about how man has
perceived light through many past generations.

Through modern advanced optics and quantum physics, we have learned
more about light in the past century than in any other period of history. The
progress is astounding, and yet light and its unconquerable enigmas remain a
tremendous puzzle. Modern science has given us access to worlds far smaller
and larger than our everyday experience would allow. But light is too small.
We chase it beyond the bounds of our scientific tools. And light is too big,
infinite in its unreachable limits. It remains elusive at both extremes.

> Some things you miss because they're so tiny you overlook them.
> But some things you don't see because they're so *huge.* We were
> both looking at the same thing, seeing the same thing, talking about
> the same thing, thinking about the same thing, except he was
> looking, seeing, talking and thinking from a completely different
> *dimension.* (Pirsig 67)

The above quote from Robert M. Pirsig, philosopher and author of the hugely successful *Zen and the Art of Motorcycle Maintenance,* encapsulates all the motivation for the writing of this book. I stop short of claiming the type of revelation and inspiration experienced by those already quoted; however, the idea for the current text and topic came to me while reading Augustine's *Confessions.* In his critique of Manichaeism, the fifth century Bishop of Hippo claimed that his old sect's refusal to believe in the Old Testament of the Bible was due to its inability to appreciate the ancient perspective from which the authors were writing. Further, in his exposition on time, Augustine employed descriptions limited to his own antiquated perspective, but expressed brilliantly an understanding of time, infinity, eternity, and the present that would rival any philosopher of our day.

It immediately became apparent to me that the ancients had contributed much to and still had valuable input on our modern scientific, philosophic, and theological understandings. I thought, perhaps light, as we understand it now and as they understood it throughout many centuries, could be a mountain at which we have all been staring, which we have all been studying and considering, and of which we have all been writing for at least three millennia. Though describing the same phenomenon, we have been describing within our limited scopes differing perspectives that catch only one view of one of the various ridgelines of the enormous mountain of light.

What would happen if someone tried to collect, synthesize, and present these differing perspectives as one coherent whole? Such an effort, that which is presented in the balance of this text, would quite aptly use light as the very window through which we might catch a passing glimpse of the most profound realities of our universe.

ADVISORIES ON THE TEXT

I must refrain from delving too deeply into the philosophy of *knowing,* but at this point I would like to attempt to make one thing clear: perfect understanding is impossible. True understanding can only come about through pure experience, and experience necessarily comes to us through our person, be it physical or psychological, and is thus perverted by our limited and biased perspective.

Then, even if we could miraculously experience a thing as it is in itself, to remove the thing from the privacy of our own knowledge, we would have to share it with others. Through what means? Best, by introducing them to the same 'objective' experience. If that be too difficult, then we might try explaining it. We could draw it or pantomime the thing, but due to our own inabilities, the likeness is bound to become very distorted. We defer to sharing our knowledge through speech or the written word. At this point, however, we must ask, 'What is a word?'

A word is a symbol that represents an agreed upon thing. And our perspectives of that thing (this book, for example) probably differ from person to person, but at least this book is this book, and we both know it. But perhaps what defines my understanding of this book or books generally relies upon one quality and yours another. Therein lies a significant problem and all the more significant as the thing we are discussing becomes more complex, less obvious, and less tangible.

This is to say simply that all language is metaphor. And unfortunately, we don't have much better means of communication. Scientists try to rectify the issue of inexact subjectivity by constructing replicable studies so their peers may review the science with their own results. Still, the formation of any duplicate study must necessarily be communicated by some means: written reports, generally. And duplicating a study purely as it was originally performed is extremely difficult.

It may be said that the mathematics scientists use is a truly universal language with little or no subjectivity, but applying the math requires language and concepts that are never as objective as we would like. No matter how diligently we try to remove the faults of subjectivity and communication, we simply cannot. We must use language to describe the occurrence and our personal experience of all physical phenomena.

To complicate matters further, we use metaphor to understand some physical phenomena in themselves. The scientific model we use to explain a complex theoretical construct does not necessarily describe the actual physical apparatus. In fact, it is possible that such models do not approach the physical reality, though they may be useful mathematical conjectures. In the words of Lewis:

> What they do when they want to explain the atom, or something of that sort, is to give you a description out of which you can make a mental picture. But then they warn you that this picture is not what the scientists actually believe. What the scientists believe is a mathematical formula. The pictures are there only to help you to

understand the formula. They are not really true in the way the formula is; they do not give you the real thing but only something more or less like it. They are only meant to help, and if they do not help you can drop them. (Lewis, Mere Christianity 37-38)

So the models of theoretical concepts, like Bohr's atom, as we learned in high school, are not necessarily true in themselves, but helpful visual metaphors? Yes. And, though most of us cannot begin to grasp the underlying mathematical formulae, we would do well to take the scientist at her word and believe the pictures she has drawn for us. Again, without getting too deep into the nature of understanding, this is all to emphasize what John Locke proposed in his *Essay Concerning Human Understanding*:

The necessity of believing without knowledge, nay, often upon very slight grounds, in this fleeting state of action and blindness we are in, should make us more busy and careful to inform ourselves than to restrain others... There is reason to think, that if men were better instructed themselves, they would be less imposing on others. (Russell 609)

In my research, I have tried to do as much, and in this text, I do not wish to impose any dogma on the reader. I believe that in as much as the world has an origin, it was impregnated with the character of the mysterious energies that formed it. Naturally, all of the descriptions of the empirical information that we gather and analyze in such a world will necessarily be indirect descriptions of that mystical character. If we find that our empirical descriptions and our concept of the divine are misaligned, we must defer to the realities of our collective experience to settle the issue. Furthermore, if we find overwhelming correlative congruencies between the two, we must not be surprised.

The world is a way. If our universe has an origin, the physical ought to bear the signature of that ultimate source.

I would like to now reflect on the feelings of the Apostle Paul in his second letter to Timothy: "Have nothing to do with foolish, ignorant controversies; you know that they breed quarrels" (2 Tim. 2:23). My theory as expounded in the following pages is yet one more metaphor by which you might understand the realities of our world. If it is not useful, rid yourself of it or file it away as just another dubious opinion. My words cannot carry the

INTRODUCTION

same weight as scripture. My arguments are of a different sort than those presented by the world's greatest theoretical physicists. This book should do little to engender controversy. It is metaphor, as is most everything we experience and know.

Here the words of the seventeenth century philosopher Benedict Spinoza are of value: "But as to what God, or the Exemplar of the true life, may be, whether fire, or spirit, or **light**, or thought, or what not, this, I say, has nothing to do with faith" [bolded for emphasis] (Spinoza 107). Agreement with or disapproval of the ideas outlined in this book has very, very little to do with one's faith or character. If my particular theory of light is helpful, hold onto it, consider it, and make the ideas your own. If not, dispose of it immediately. Intellectual assent to the model has little to do with the underlying 'formula'. I do not claim to know precisely what that divine formula is, but its model is becoming clearer the more we understand its physical expression in our world. Consider only what you can see, and you may begin to understand the 'unknown but still reasoning thing' behind that.

I leave this introduction with a reminder from scripture for a portion of my readers, believers at the outset: "You are a chosen race, a royal priesthood, a holy nation, a people for his own possession, that you may proclaim the excellencies of him who called you out of darkness into his marvelous light" (1 Peter 2:9).

And to all, regardless of your creeds, consider the truth that will be elucidated in the coming pages: somehow a thing that is not really a thing is the most important thing, the only thing, through which everything attains reality. Time and space cannot exist without this thing, and it is the only thing we have ever known and experienced. Light is the crux of it all.

MARVELOUS LIGHT

EARS TO HEAR

Nature and Nature's laws lay hid in night.
God said, 'Let Newton be,' and all was light.
- Alexander Pope -

If I have seen further, it is by standing on the shoulders of giants.
- Isaac Newton -

Science. Philosophy. Theology. Imposing topics of discussion and formidable areas of study, yet much of the human race is drawn to these ideas, regardless of the individual's education. Few, if asked, would not willingly share his or her opinion (no matter how informed) on any of the three disciplines, and likely, conversation started, passionate debate would ensue. Many families and friends have learned to avoid deeper topics like these to safeguard civility.

In part, the contentiousness of these topics is rooted in man's belief that there is something underlying the reality of our daily experience. We long to find a grand theory, a perfect reasoning, a divine character, but it is never as clearly manifest as we need it to be to convince our counterpart in debate:

> What is eternal and important is often hidden from a man by an impenetrable veil. He knows: there's something under there, but he cannot *see* it. The veil reflects the daylight. (Wittgenstein, Culture and Value 80e)

Unfortunately, the greatest truths in our world are disguised by the inescapable mirror of the mundane. Science must speak of the mirror and its reflection only, for that is what demonstrates the world's phenomena. Religion speaks of what is unseen on the other side of the mirror. And philosophy finds itself in a tricky spot:

> All *definite* knowledge – so I should contend – belongs to science; all *dogma* as to what surpasses definite knowledge belongs to

11

theology. But between theology and science there is a No Man's Land, exposed to attack from both sides; this No Man's Land is philosophy. (Russell xiii)

In order to appreciate the complexities of the world as we know it, it is important to engage philosophy to act as a bridge between the science we know tangibly and the spirituality we believe inherently. To lift a corner of the 'impenetrable veil', we must employ all human knowledge in tandem. To see further, we must not neglect the giants' shoulders available to us.

In pursuing the impossible task of summarizing Christian theology, Augustine comments:

Unless I am much mistaken, this argument will be more difficult and will require discussion of great subtlety. We shall have to engage with philosophers, and philosophers of no ordinary sort, but those who enjoy the most eminent reputation amongst our adversaries. (Augustine, City of God 46)

Augustine had the humility to recognize that a discussion of any worth on such an enormous topic would have to engage with and comment on the greatest philosophies of his time, even ideas from philosophers whose opinions contradicted his own. Likewise, today, in order to speak at length and in great breadth on any given topic, one must have a thorough understanding of the arguments of his friends and foes alike.

"The real train of knowledge isn't a static entity that can be stopped and subdivided" (Pirsig 362). The collection of human understanding is dynamic and constantly progressing. To keep up, one needs to keep a broad view. He cannot only rely on himself or his one area of study. Surely, he must trust his experience but cannot trust this alone.

Since we can only think by means of ideas, and since all ideas come from experience, it is evident that none of our knowledge can antedate experience. (Russell 610)

Unfortunately for the individual thinker, many ideas we must receive from others: one man cannot possibly experience all himself. Much of our knowledge is attained on the basis of faith, faith in those who have experienced and seen what we now believe. Our own reason aids us in understanding the lines of argument presented by scientists, theologians, and philosophers, but the information that is shared with us is often not from our own experience. We do this, hardly considering the leap of faith it takes. We know that the reasonings of others:

> ...do not come from us and at the same time we recognize in them, *because of their harmony*, the work of reasonable beings like ourselves. And as these reasonings appear to fit the world of our sensations, we think we may infer that these reasonable beings have seen the same thing as we; thus it is that we know we haven't been dreaming. (Pirsig 343)

We must rely on the experience and expertise of others, not only to broaden our knowledge, but also to ensure it. The subjectivities of our own thoughts melt away a bit when they are supported by the arguments of others. The less information we allow from outside ourselves, the less ability we have to establish our own ideas. We need science to ensure what we know and to explore what we do not. Likewise, we need philosophy, and importantly, we cannot belittle the contribution of theology.

> The ordinary man believes in the Solar System, atoms, evolution, and the circulation of blood on authority – because the scientists say so. Every historical statement in the world is believed on authority... A man who jibbed at authority in other things as some people do in religion would have to be content to know nothing all his life. (Lewis, Mere Christianity 41)

It is so easy for the modern academic to recognize the generalized importance of understanding or deferring the final judgement to science. But the balance of wisdom from all of the world's great sages cannot be ignored, including the philosophical and theological. These are of great importance, and every man must determine for himself what steps must be taken to overcome his own natural prejudices and preferences in order to accept a broader array of human wisdom. We must have the humility to accept that our intellectual foes may have insightful knowledge in and outside of our area of expertise. We must listen to each other.

CAVEAT ON CHRISTIANITY

Because much of what we read in western academia is from western thinkers, naturally many of the critiques on religion are direct critiques on Christianity; however, these western arguments often apply, and more compellingly, to eastern religions. For now, however, we will stick to west.

> The less men know of nature the more easily can they coin fictitious ideas, such as trees speaking, men instantly changing into stones, or into fountains, ghosts appearing in mirrors, something issuing from nothing, even gods changed into beasts and men, and infinite other absurdities of the same kind. (Spinoza 234)

Spinoza, a self-styled believer but more accurately regarded as a secularist, spoke the above words in exhortation to religious people to rid themselves of beliefs that contradicted their natural experience. He felt that knowledge of physical laws would make clear the impossibilities contained in scripture so that the devout might focus on the more important moral lessons of the Bible. He wanted them to balance their spiritual belief with reason, but in the end, Spinoza relied totally on natural science. We might find an actual balance between science and theology in the thoughts of Bertrand Russell, one of the twentieth century's preeminent philosophers:

> Science tells us what we can know, but what we can know is little, and if we forget how much we cannot know we become insensitive to many things of very great importance. Theology, on the other hand, induces a dogmatic belief that we have knowledge where in fact we have ignorance, and by doing so generates a kind of impertinent insolence towards the universe. Uncertainty, in the presence of vivid hopes and fears, is painful, but must be endured if we wish to live without the support of comforting fairy tales. (Russell xiv)

Doubt is uncomfortable, and most people seek to rid themselves of doubt whenever possible. But whenever nothing of certainty can be assured, doubt is the only reasonable course to take. The secularist can take this to extremes. Scientists often wish for the religiously-oriented to only give intellectual credence to the scientifically demonstrable and to abandon any definitive testimony of scripture. Carl Sagan, a twentieth century astrophysicist, public figure of science, and fierce detractor of religion and philosophy, excoriated the comfort we seek in the face of doubt in saying:

> What do we really want from philosophy and religion? Palliatives? Therapy? Comfort? Do we want reassuring fables or an understanding of our actual circumstances? Dismay that the Universe does not conform to our preferences seems childish. You might think that grown-ups would be ashamed to put such disappointments into print. The fashionable way of doing this is not to blame the Universe – which seems truly pointless – but rather to blame the means by which we know the Universe, namely science. (Sagan 46)

Sagan shares these sentiments against philosophy and religion with Galileo, who described prioritization of a particular type of scientific reason over sensory perception:

> Nor can I ever sufficiently admire [Copernicus and his followers]; they have through sheer force of intellect done such violence to their own sense as to prefer what reason told them over what sensible experience plainly showed them. (Sagan 39)

In Copernicus and Galileo's time, when direct visual confirmation of a spinning globe could not be attained by NASA imagery, mathematical reason had to reign over terrestrial sense for scientists to believe the heliocentric model of the solar system. Those ancient scientists ignored old interpretations of scripture that lent credence to a geocentric model. They found an important and valid authority outside of scripture, yet Sagan glorifies the authority of science beyond what it can bear. Sagan's disdain for the other major disciplines of human knowledge cannot simply be overlooked. His passionate hatred for the real and perceived negative effects religion has had on human history is palpable.

> Philosophy and religion presented mere opinion – opinion that might be overturned by observation and experiment – as certainty. This worried them not at all. That some of their deeply held beliefs might turn out to be mistakes was a possibility hardly considered. Doctrinal humility was to be practiced by others. (Sagan 13)

Of course, no such problem arises for those many religious people who treat the Bible and the Qur'an as historical and moral guides and great literature, but who recognize that the perspective of these scriptures on the natural world reflects the rudimentary science of the time in which they were written. (Sagan 24-25)

While Sagan does raise valid points that need to be and will be addressed, his vitriol detracts notably from the message he was trying to convey. By lumping together all men and women who believe that scripture is more than great literature, he readily disposed of the valid philosophies of many of history's greatest thinkers. To deny the wisdom and understanding of these men is to open oneself to criticism of the highest degree. Sagan's unfettered passion should give us pause when we consider what he is actually proposing.

In discussing the unswerving beliefs of the religious, others, I believe, comment interestingly on Sagan's thoughts expressed above:

When people are fanatically dedicated to political or religious faiths or any other kinds of dogmas or goals, it's always because these dogmas or goals are in doubt. (Pirsig 190)

For it is an observed fact that men employ their reason to defend conclusions arrived at by reason, but conclusions arrived at by the passions are defended by the passions. (Spinoza 57)

I would not make the claim that Sagan doubted his dogma of science, though he was likely unsure of the end results of his own theories, but I certainly will say that what he was defending was dogma, as much as any religious belief. For him, science was all. It would, in time, explain all. There was no need to study any other subject. Science was his god, and of course, he excelled in his worship far more than all but a tiny sliver of humanity. He used reason to defend a great portion of his ideas, but where his passion prevailed, his argument became disjointed, as was his criticism of religion.

What Sagan failed to see was that with the same passion as he pursued science, others could pursue religion without doing injustice to their own intelligence or the intelligence of others. Augustine refers to the importance of using his reason when discussing the Manichean faith:

In regard to the physical world and all the natural order accessible to the bodily sense, consideration and comparison more and more convinced me that numerous philosophers held opinions much more probable than theirs. Accordingly, after the manner of the Academics, as popularly understood, I doubted everything, and in the fluctuating state of total suspense of judgement I decided I must leave the Manichees, thinking at that period of my skepticism that I should not remain a member of a sect to which I was now preferring certain philosophers. (Augustine, Confessions 89)

Through his intellect, Augustine was able to rid himself of the doctrines of the religious sect from which he received his education. In his understanding of the physical world, he refused to accept theories that contradicted reality. A supremely religious man, Augustine used his brain. The condescension of Sagan falls flat against the brilliance of Augustine and the theorizing of many of today's most ardent religious thinkers.

C.S. Lewis suggests to his audience, "I am not asking anyone to accept Christianity if his best reasoning tells him that the weight of the evidence is against it. That is not the point at which Faith comes in" (Lewis, Mere Christianity 78). Many earnest Christians would eagerly reject the concept of blind faith. Religious conviction must be tempered by reason and experience. If nothing else would convince my secular reader of what Sagan could not accept, let the arguments in this book stand as an example of religion, science, and reason coexisting peaceably.

In a section entitled 'Caveat on Christianity', there has been very little critiquing of the religion, so let us give credit to Sagan where credit is due. What irritated him perhaps more than anything was the Catholic Church's refusal to recognize Galileo's verification that the earth revolved around the sun and its active attempt to bury the evidence and reason that supported Galileo's conclusion. Sagan saw the medieval polity of the Church as the antithesis of free and productive inquiry. Truly, it was. The efforts of the Church purposefully set science back for centuries throughout the Middle Ages and beyond because religious leaders did not possess the intellectual imagination to rethink their stance on various scriptural topics. By reevaluating their interpretation of scripture, these men would have readily found that the advances in science contradicted none of their divine story, as the modern Christian must now understand.

After the Reformation, churches in Protestant countries had far less influence than did the Catholic Church in its geographical and political havens. "This was wholly a gain, for the Churches, everywhere, opposed as long as they could practically every innovation that made for an increase of happiness or knowledge here on earth" (Russell 529). Only after religion lost its iron grip on western culture was science able to flourish without impediment and did organized Christianity gain some flexibility to allow scientific and theological understanding to progress side by side. With the advent of Protestantism and the separation of church and state, religious leaders were freed of the temptation toward political power and more inclined toward free inquiry within the sciences.

That is not to say that everything has been smooth sailing since then. In conservative, rural America especially, we still have that old ire of the church that has since inspired the Scopes Monkey Trial and many similar public debates of science v. religion. Religious folks still make the errors of the medieval church, and scientific folks continue to make the mistakes of Sagan. Neither has the stomach to recognize the bounds of their own discipline.

FENCES

It is difficult to emphasize how important it is for everyone to recognize the limits of his or her own intelligence. We must possess enough humility to understand where we lack insight, and at those times, we must defer to the more learned, the wise. The overlap between areas of study only goes so far, and it is vital for those within any given academic discipline to recognize the fences beyond which their discipline cannot make any valid comment.

The authority of science, which is recognized by most philosophers of the modern epoch, is a very different thing from the authority of the Church, since it is intellectual, not governmental... It prevails solely by its intrinsic appeal to reason. It is, moreover, a piecemeal and partial authority; it does not, like the body of Catholic dogma, lay down a complete system, covering human morality, human hopes, and the past and future history of the universe. It pronounces only on whatever, at the time, appears to have been scientifically ascertained, which is a small island in an ocean of nescience... the pronouncements of science are made tentatively, on a basis of probability, and are regarded as liable to modification. (Russell 492)

Scientific knowledge is, then, a state of capacity to demonstrate... for it is when a man believes in a certain way and the starting-points are known to him that he has scientific knowledge, since if they are not better known to him than the conclusion, he will have his knowledge only incidentally. (Aristotle 105)

As much as scientists may not like to hear it, the honest among them must admit that science has set, defined limits. Scientific knowledge necessarily rests on the observable, the systematic, the duplicable. Science must start with known criteria in order to have a basis on which to test unknown variables. It must have a control. It also must have a means of measurement before it can measure, an oddly simultaneous activity.

So – you say – *what* is judged here is independent of the method of judging it. What length *is* cannot be defined by the method of determining length. – To think like this is to make a mistake... What "determining the length" means is not learned by learning what *length* and *determining* are: the meaning of the word "length" is learnt by learning, among other things, what it is to determine length. (Wittgenstein, Philosophical Investigations 225e)

Everything scientific must be observed and defined by predetermined means, but there are no predeterminations at the outset. The process is tricky. It is active. The theoretical beginnings of science are not exactly science; however, modern science has such a stalwart base of predeterminations, observations, measurements, and objective facts that for practical purposes we can discuss 'pure science'. But always, assumptions must be made, and only upon these unscientific assumptions can scientific inquiry begin. Philosophy becomes important for the scientist when he realizes:

The scientist can have no recourse above or beyond what he sees with his eyes and instruments. If there was some higher authority by recourse to which his vision might be shown to have shifted, then that authority would itself become the source of his data. (Kuhn 114)

Following this logic, if not in any other way, philosophy enters into the domain of science at least in the inspiration of scientific hypotheses. "One might also give the name 'philosophy' to what is possible *before* all new discoveries and inventions" (Wittgenstein, Philosophical Investigations 50e). That divine spark, the ethereal something which comes to the theoretical scientist in his most prophetic moments, the formation of hypotheses, cannot be science. It is not observed, nor observable. It is not tangible, nor tangibly expressed. Its mystery is not one of reason and enigma, but of utter inscrutability. How exactly is the scientist supposed to deal with that?

> Galileo had a far-reaching answer: Science should deal only with those matters that can be demonstrated. Intuition and authority have no standing in science. *The* only *criterion for judgement in science is experimental demonstration.* (Rosenblum and Kuttner 26)

This definition is far stricter than the modern audience will typically allow. Many of the topics to be discussed in this very book cannot necessarily be said to have been demonstrated experimentally. Those that we may accept as scientific certainty have been verified by mathematical demonstration and advanced theoretical computation, as opposed to true experiment, in some cases. With that said, modern science has greatly benefitted from the creativity of its practitioners in discovering new ways to indirectly access the far reaches of our micro and cosmic-worlds. We are constantly pushing the bounds of the physically observable in an attempt to confirm what theorists have caught glimpses of beyond their fence. Quantum physicists, as we will see, are among the fortuitous scientific philosophers who are able to pursue their own work to realms that have implications well beyond their area of study and beyond their capacity to directly demonstrate.

Galileo's restrictive stance, stated above, has perhaps been forgotten by today's boldest voices in science, those for whom compelling yet unconfirmed theory is enough to ensure that science is not far from explaining every dot and tittle in our expansive universe and beyond. It is with this boldness that Neil DeGrasse Tyson can say, "The good thing about science is that it's true whether or not you believe in it," which he has said on *Real Time with Bill Maher* and Twitter, if not on various other occasions. In today's internet age where soundbites carry more weight than fully reasoned arguments, Tyson's words are plastered ubiquitously on memes, bumper stickers, and the like. We do well to dissect the quote before moving on.

First, by saying that science is true (with no qualifications) is to accept the concept of truth, which many of Tyson's contemporaries are abandoning. He seems to be certain of truth in the same ways for which religious folks are mocked. Eighteenth century philosopher David Hume felt that no amount of empirical evidence, however constant and compelling, was ever grounds for us to establish true cause and effect by means of reason. Reason will only ever allow us the probability of correlation. Where no direct relationships can be established, truth becomes very slippery indeed. Furthermore:

> Karl Popper, in his 1934 book *The Logic of Scientific Discovery*, enthusiastically dispatched the ambition of the logical positivism that theories could be proved true, and introduced the now commonplace notion that it is only possible to prove theories false. Theories became more credible, he argued, the more tests they pass, but no matter how well they do, they remain always vulnerable to disproof by some novel experiment. Theories can never gain any guarantee of correctness. Science builds up an increasingly complete picture of nature, but even the most treasured laws of science remain subject to repeal, should the evidence demand it. (Lindley 206)

So, Tyson's 'truth' statement flies in the face of modern scientific understanding, but perhaps more importantly, we now have to understand what he means by 'science'. If Tyson is saying, 'All physical phenomena are true in themselves and have valid causes,' I would heartily agree with him. If he is saying, 'All scientific observation describes with reasonable accuracy anecdotal incidents and contributes to a systematic understanding of the physical world,' again I would agree with him. If Tyson is recognizing the clear limits of scientific verifiability, I would be on his side, but something tells me that this is not what he is saying. Something tells me that by employing the inexact qualities of rhetoric in conversation, Tyson is making a bolder statement than he would ever support in earnest debate. Tyson knows better than I the limits of science, yet this quote makes current scientific understanding seem flawless. History and today's state of science say something very different. Scientific data may be extremely precise, but the conclusions drawn from the data are always less certain.

In the flurry of scientific theory that characterized the early 1900s, Ernst Mach felt that "a theory was merely a set of mathematical relationships linking tangible phenomena" (Lindley 63). This understanding of quantum physics and Einsteinian relativity is far more limiting than Tyson's quote. If theory is only a mathematical construct that quantifies the relationship between physical phenomena, though the consequent equations may be of great use and impressive predictive quality, science's 'truth' would only be expressible through its usefulness. This is a supremely pragmatic point of view, and one with little to do with truth. Brian Greene, a modern theoretical physicist who has worked to popularize string theory, notes:

> The history of science teaches us that each time we think that we have it all figured out, nature has a radical surprise in store for us that requires significant and sometimes drastic changes in how we think the world works. (Greene 373)

Time and time again throughout the entirety of history, sometimes falsely and sometimes progressively, new scientific discovery has proven old scientific understanding wrong. That is not a knock on science. In fact, that is science doing precisely what science is designed to do. By eliminating falsifiable theories, scientists are constantly refining and 'perfecting' our understanding of the physical world. But all that we can definitively say about these new theories is that they proved the old theories false. One experiment might shut the door to an old interpretation while at the same time opening the window to various new possibilities. Scientific inquiry ultimately asks far more questions than it answers.

> If the purpose of scientific method is to select from among a multitude of hypotheses, and if the number of hypotheses grows faster than experimental method can handle, then it is clear that all hypotheses can never be tested. If all hypotheses cannot be tested, then the results of any experiment are inconclusive and the entire scientific method falls short of its goals of establishing proven knowledge... Scientific truth [is] not dogma, good for eternity, but a temporal quantitative entity that [can] be studied like anything else. (Pirsig 140)

Through multiplication upon multiplication of facts, information, theories and hypotheses, it is science itself that is leading mankind from single absolute truths to multiple, indeterminate, relative ones. The major producer of social chaos, the indeterminacy of thought and values that rational knowledge is supposed to eliminate, is none other than science itself. (Pirsig 142)

Science, though telling us which interpretations of physical phenomena we ought to dispose, also leaves the door open due to its persistent refusal to claim anything as positive and absolute. It *ought* to do nothing else, but that does not make the scientist's job any easier or science's implications any simpler. And it does little to support Tyson's quote. We do best to rid ourselves of that kind of scientific absolutism. And still, we should never fail to appreciate the progress science has made and how it has contributed to a far broader base of human knowledge century over century, and increasingly, year over year.

To the philosophical limits of science, there is a knee-jerk reaction that one might easily fall into: scientific materialism. Scientific materialism supposes that only physical things composed of matter and energy are real. Anything else is not. If the scientist cannot take a measurement of any given phenomenon, then that thing does not properly exist. Thoughts and emotions are only as real as the neurons and chemicals communicating them, and consciousness must have a very limited definition. If anything in our universe cannot be quantified and ultimately studied physically, the scientific materialist does not believe that it exists in any real way.

Of course, extreme materialism is a philosophy of science that may be sound in theory but is all but impossible to defend in practice. It is more accepted among the lay followers of popular science as opposed to the actual scientists themselves. Every man, the religious and the secularist, is bound to believe in many things that lie on the other side of the boundaries of materialism, things that cannot be so easily rooted back into the physical world. Scientific materialism ebbs and flows in popularity, contradicted by spiritualists and the religiously devout. It is a tenuous philosophy of science, but one that enjoys lengths of time in the cultural limelight.

Against materialism, detractors can easily observe:

When the man of the five senses talks of *truth*, he regards it but as a predicate of something historical or scientific proved a fact; or, if he allows that, for aught he knows, there may be some higher truth, yet, as he cannot obtain proof of it from without, he acts if under no conceivable obligation to seek any other satisfaction concerning it. Whatever appeal be made to the highest region of his nature, such

23

a one behaves as if it were the part of a wise man to pay it no heed, because it does not come within the scope of the lower powers of that nature. (MacDonald 226)

The ornery critic of this point of view can go on in the most facetious style:

The laws of science contain no matter and have no energy either and therefore do not exist except in the people's minds. It's best to be completely scientific about the whole thing and refuse to believe in either ghosts or the laws of science. That way you're safe. That doesn't leave you very much to believe in, but that's scientific too. (Pirsig 39)

The reality is that as important and comprehensive as science is in our world, there are some things science simply cannot say but on which man still wants comment. Purpose, ultimate cause, origins; science might be able to theorize about these topics, but no truly scientific statements can be written on them.

When considering the *why* behind the world, the scientist must give up in the end. Why the world is the way it is, why the world *is* at all, and whether or not there is something behind all the physicality we observe, these are questions that science cannot begin to answer. The statements that there is nothing more in the world beyond the physicality we observe or that there is something profound, perhaps spiritual, guiding our lives, neither of these are things that science can say. Necessarily, if something is or is not beyond the physical world, we cannot comment on that possibility with a study limited to that physical world.

Modern science is an amazing accomplishment of human knowledge, but it still only has the ability to comment on actual physical phenomena. Scientific truths are simply scientific facts; infinitesimally small data points in an expansive world of experience. Considering science's limits:

The word *truth* ought to be kept for higher things. There are those that think [scientific] facts the highest that can be known; they put therefore the highest word they know to the highest thing they know, and call the facts of nature truths; but to me it seems that, however high you come in your generalization, however wide you make your law – including, for instance, all solidity under the law of freezing – you have not risen higher than the statement that such and such is an invariable fact. Call it a law if you will – a law of nature if you choose – that it always is so, but not a truth. It cannot be to us a truth until we descry the reason of its existence, its relation to mind and intent, yea to self-existence. Tell us why it *must* be so, and you state a truth. (MacDonald 227)

George MacDonald, a nineteenth century theologian who penned the above quote, went on to focus on his topic of discussion, the Christian God, and again remarked on science's limits.

Human science cannot discover God; for human science is but the backward undoing of the tapestry-web of God's science, works with its back to him, and is always leaving him – his intent, that is, his perfected work – behind it, always going farther and farther away from the point where his work culminates in revelation... science will never find the face of God. (MacDonald 228)

It is very important to note at this juncture that science is hardly to blame for its limits. As a discipline, science must stand stalwart in its purpose to describe and analyze physical phenomena, and nothing else. Were it to betray that mission, it would cease to be science. Since this form of study cannot be to blame for its limits, the theologian and philosopher must not impose impossible requirements on the scientist. Science tells us what we can know physically and must stop far short of answering man's oldest and, often, most pressing questions.

Lastly, to the unphilosophical person who would find shelter in the theories of scientific materialism, denying the existence and importance of other areas of study, I have not much to say, for of course any argument I will present must necessarily encapsulate words and ideas outside of the materialism that he accepts as his totality. I would have to ask that he learn to humbly hear the arguments of those considered to be the wisest men of history. If we cannot listen to the voices of those outside of our fence, we are guaranteed a very narrow view of the world indeed.

LISTENING

It is the highest ignorance and pride that allows a man of any discipline to wholly disregard the truths illuminated by areas of study outside of his domain. Where we have not the learning to speak, we must be silent and have ears to hear experts of other fields.

Science, philosophy, and theology have plenty to say on various topics, many that touch all three systems of thought. Let us briefly consider *purpose*. Purpose is a readily found concept in theology and philosophy, but in physics it seems not to have a place; however, as Niels Bohr pointed out, "the concept of purpose, which is foreign to mechanical analysis, finds certain application in biology" (Lindley 202). So, a concept that finds a home in all three major realms of thought is still barred from application in a subset of scientific topics. A student of any field will find at some point that his discipline cannot comment on a particular topic, and at that time, he too should remain silent and ready to hear those which may.

"A clock made with wheels and weights observes all the laws of nature just as precisely when it is made poorly and fails to show the correct time as when it satisfies the artisan's intentions in every respect" (Descartes 66). So, the artist and physicist both have reason to comment on the clock's workings, each from very unique perspectives. People are here tempted, however, to segregate these comments and label them according to the commenter.

> People nowadays think that scientists exist to instruct them, poets, musicians, etc. to give them pleasure. The idea *that these have something to teach them* – that does not occur to them. (Wittgenstein, Culture and Value 36e)

Perhaps this opinion is held for good reason. Is it the professionals themselves who are to blame for likewise disregarding the other modes of knowledge that influence our world?

> We have artists with no scientific knowledge and scientists with no artistic knowledge and both with no spiritual sense of gravity at all, and the results is not just bad, it is ghastly. (Pirsig 377)

This was not always the case. The 'Renaissance Man' is a prime example of the capacity of men and women to gain expertise in various studies. Leonardo da Vinci was such a man, as have been many of the clergy throughout the centuries, east and west.

> Jesuits and many other churchmen had a surprisingly broad and rigorous education in philosophy, logic, and even mathematics. Such men were singularly well equipped to deal with the problems that we might now call cross-disciplinary. (Lindley 16)

Within any religion, scientifically educated clergy are valuable. Their religious importance is very well summed up in a quote from Spinoza:

> Further, since without God nothing can exist or be conceived, it is evident that all natural phenomena involve and express the conception of God as far as their essence and perfection extend, so that we have a greater and more perfect knowledge of God in proportion to our knowledge of natural phenomena; conversely (since the knowledge of an effect through its cause is the same thing as the knowledge of a particular property of the cause) the greater our knowledge of natural phenomena, the more perfect is our knowledge of the essence of God (which is the cause of all things). (Spinoza 34)

If there is a more compelling reason for the Christian or devotee of any religion to study science, it is unknown to me. In introducing the Philosophy of Illumination, a twelfth century Islamic philosophy, Suhrawardī compares this respect for scientific observation with divine revelation:

> Indeed, the system of the Illuminationists cannot be constructed without recourse to luminous inspirations, for some of their principles are based on such lights... Just as by beholding sensible things we attain certain knowledge about some of their states and are thereby able to construct valid sciences like astronomy, likewise we observe certain spiritual things and subsequently base divine sciences upon them. He who does not follow this way knows nothing of philosophy and will be a plaything in the hands of doubts... Whoever wishes to learn the details of this science –

which is merely a tool – should consult the more detailed books. (Suhrawardi 4)

Suhrawardī was apparently at ease working between and comparing science and philosophy, a philosophy of his that was built directly upon theology, while also referring his readers to books of greater authority for those topics they wished to pursue in greater detail. Such comfort with cross-disciplinary thought has been dwindling from popular culture, much to the detriment of general knowledge and academic practice. At the beginning of the twentieth century, however, Einstein gave academia and culture a much-needed shot in the arm:

> Perhaps no idea in the whole history of physics can match the record of relativity for creating a widespread public stir while being understood by practically nobody... For since the day of Einstein there has been a growing gulf between the world of science and the world of letters. Almost by accident, Einstein the man and his works were one of the few bridges spanning the gulf. (March 159)

The 'spooky' aspects of modern theoretical physics, like Einsteinian relativity, are very attractive to the religious opportunist who believes that in the generally accepted enigmas of science he can more easily use physical evidence to support his theological ideas. As attractive as this option may be, modern academics preempt the theologians by saying,

> It is sometimes implied that the sages of ancient religions intuited aspects of contemporary physics. The argument can go on to claim that quantum mechanics provides evidence for the validity of these mystic teachings. Such reasoning is not compelling. (Rosenblum and Kuttner 251)

One could easily claim that this book is as opportunistic as those sources referred to in the quote, but one important qualification must be made: I do not believe that the fathers of ancient religions had some divine insight into the mechanical workings of our physical world. Their scientific knowledge, as Sagan knew, was as rudimentary as that of their contemporaries. But what this book does contend is that our religious, philosophic, and scientific forefathers were describing the same reality we are still describing today. Insofar as we are looking at the same thing from very different perspectives, our theories, if generalized, should align fairly well, though our words may not.

> A theology which insists on the use of *certain particular* words and phrases, and outlaws others, does not make anything clearer... It gesticulates with words, as one might say, because it wants to say something and does not know how to express it. *Practice* gives the words their sense. (Wittgenstein, Culture and Value 85e)

The importance of *practice* and the *present* will be made specifically and abundantly clear in later chapters; however, for now we must keep our generalized view of the disciplines we are discussing. Science works with details and builds up from the most basic pieces of information. Philosophy and theology, on the other hand, must attempt to get their hands around much girthier ideas: "Science works with chunks and bits and pieces of things with the continuity presumed, and [unscientific study] works only with the continuities of things with the chunks and bit and pieces presumed" (Pirsig 205). Comparing science with theology and philosophy is difficult, if not impossible, as is comparing scientists, theologians, and philosophers. They are very different practices and very different practitioners. And by limiting the 'certain particular words' that we are allowed to use, intellectual authorities are also limiting the connections that otherwise can be made quite readily through lingual plasticity. In all areas of study, however:

> Two things are to be remembered: that a man whose opinions and theories are worth studying may be presumed to have had some intelligence, but that no man is likely to have arrived at complete and final truth on any subject whatever. When an intelligent man expresses a view which seems to us obviously absurd, we should not attempt to prove that it is somehow true, but we should try to understand how it ever came to *seem* true. (Russell 39)

In that spirit, many different voices will be heard in the present work, voices of intellect that rarely find themselves quoted in the same text. To tackle a topic as big as *light*, with all of its complexities and its long history of study throughout human existence, competing opinions must be heard and seemingly contradictory assertions must be coalesced into one uniform whole. For this reason, as will certainly have been noticed at this point, many quotes from various sources will be utilized early and often. Many of the quotes will be from thinkers who are still considered the most respected and influential theologians of all time, including St. Augustine and C.S. Lewis. We must also consider the perspectives of other world religions, including Islam and Buddhism. Philosophers of the highest reputation, both ancient and modern, will have their chance to speak. And lastly, of course, we must engage with modern scientific thought found in the classroom and in popular literature.

This catalog of imposing opinions should make it clear that we are concerned with serious and difficult theory. In the unexpected harmonies of the many voices compiled in this text's references, a grand theory of truth is being proposed. The final thesis is guaranteed to be complex, but I beg the reader's attention, patience, and every shred of his or her intelligence. The reading will not be easy as we discuss quantum physics, Einsteinian relativity, paradoxical theology, and dense philosophy, but:

> If our religion is something objective, then we must never avert our eyes from those elements in it which seem puzzling or repellent; for it will be precisely the puzzling or the repellent which conceals what we do not yet know and need to know. (Lewis, The Weight of Glory 34)

> There is no good complaining that these statements are difficult. Christianity claims to be telling us about another world, about something behind the world we can touch and hear and see. You may think the claim false, but if it were true, what it tells us would be bound to be difficult – at least as difficult as modern Physics, and for the same reason. (Lewis, Mere Christianity 86)

CONVERGENCE

If individual fields of human knowledge are able to say that they are studying things as they really are, as each one refines its knowledge and approaches more and more closely the ultimate truth of its field, each would approach more and more closely THE ultimate truth. With this kind of inward concentric movement, we ought not be surprised to find that each discipline approaches the others at the very center of reality. Insofar as each is legitimate in itself, this convergence must also be legitimate. We are all studying the same world, after all.

Unfortunately, those who do not believe in the legitimacy of the other fields, especially those who are militantly opposed to lending any credence to religion or science, stifle the magnetism that would naturally unite all knowledge and experience. Sagan and his defenders are of the unfortunate perspective that disrupts opportunity for intellectual harmony. Early in *Pale Blue Dot*, Sagan discusses the infamous tree of knowledge in the biblical account of the creation of man, in Genesis 2 and 3. The astrophysicist lamented God's command restricting man from the tree of knowledge saying:

Knowledge and understanding and wisdom were forbidden to us in this story. We were to be kept ignorant. But we couldn't help ourselves. We were starving for knowledge – created hungry, you may say. (Sagan 53)

He also comments:

In some respects, science has far surpassed religion in delivering awe... A religion, old or new, that stressed the magnificence of the Universe as revealed by modern science might be able to draw forth reserves of reverence and awe hardly tapped by the conventional faiths. (Sagan 50)

Unfortunately, in his book, Sagan commits two logical fallacies. First, he is too intently focused on the *slippery slope* of interpretive literalism. He is

31

very concerned that those who took some of the words of the Bible literally would be forced to accept ridiculous concepts as truth. And the idea of earnest allegorical interpretation seems not to be an option he allowed the religiously inclined (perhaps because allegory has so little application within his practice of science). Understanding the Bible in this way, he does not understand the text he references at all.

Secondly, he sets up the *straw man* of rudimentary theology and feels confident that he has disposed of the inconvenience of religion by knocking that straw man down. He argues effectively against the Church and its general practices. Unfortunately, he argues with the Church of the Middle Ages, not of his own time. Sagan might have been surprised to find that there is already at least one religion that 'stresses the magnificence of the universe as revealed by modern science'.

> Think for a moment what would be our idea of greatness, of God, of infinitude, of aspiration, if, instead of a blue, far withdrawn, light-spangled firmament, we were born and reared under a flat white ceiling! I would not be supposed to depreciate the labours of science, but I say its discoveries are unspeakably less precious than the merest gifts of Nature, those which, from morning to night, we take unthinking from her hands. (MacDonald 173)

> When I look at your heavens, the work of your fingers,
> > the moon and the stars, which you have set in place,
> what is man that you are mindful of him,
> > and the son of man that you care for him? (Ps. 8:3-4)

By all appearances, Sagan disregarded modern religion and all its appreciation for the profundities of the cosmos in his discussion of modern science. The religious too often make the opposite mistake in their staunch refusal to contend with scientific discoveries that add difficulty to their particular scriptural interpretations. To see where one's unquestioned focus lies, theological or otherwise, pay close attention to those words that he capitalizes in his text. These priorities can blur his perspective of clear truths.

"The feuding, biased, and flawed beliefs of both physics and religious teachers make it difficult for maturing minds, especially, to develop sound beliefs" (Pelton 12). To have legitimate conversation, we must recognize each other's legitimacy and not allow ourselves to lapse into complete disregard for the ideas of others. We must allow that there may yet be truth out in the world that is unknown to us but known to others, perhaps in ways very different from our knowledge. We must pursue truth wherever we find it. "The pursuit of truth, when it is whole-hearted, must ignore moral considerations; we cannot know in advance that the truth will turn out to be what is thought edifying in a given society" (Russell 78). Bertrand Russell goes on to quote John Locke in this vein of thought:

> One unerring mark of love of truth, he says, is "not entertaining any proposition with greater assurance than the proofs it is built upon will warrant." (Russell 607)

Perhaps it is easier, as mentioned before, for the religious to recognize the legitimacy of the sciences. Augustine discusses the Christian's outlook on human reason, scientific and theological:

> In matters apprehended by the mind and the reason it has most certain knowledge, even if that knowledge is of small extent... It also trusts the evidence of the senses in every matter; for the mind employs the senses through the agency of the body, and anyone who supposes that they can never be trusted is woefully mistaken. It believes also in the holy Scriptures, the old and the new, which we call canonical, whence is derived the faith which is the basis of the just man's life, the faith by which we walk on our way without doubting, in the time of our pilgrimage, in exile from the Lord. So long as this faith is sound and certain we cannot justly be reproached if we have doubts about some matters where neither sense nor reason give clear perception, where we have received no illumination from canonical Scriptures and where we have not been given information by witnesses whom it would be irrational to distrust. (Augustine, City of God 879)

The religious can, and must, trust their senses and reason and the witness of others who are known to have reliable sense and reason. The religious must accept this in ways similar to how they would trust their holy scriptures. More, where science, philosophy, or theology have yet to make comment, the religious have excuse to allow doubt to reign. It would be foolish to do otherwise. Perhaps there is no reciprocal for the scientist, who, though naturally trusting his own senses, finds no reason to trust his spiritual experience or defer to philosophical reason. The argument for these, as would be expected, is far less concrete and straightforward than the argument for the physical, yet, likely, no less legitimate and real.

Of course, this is not to say that the man of science has no experience of the spiritual. If there is a spiritual aspect in this world, he experiences it. Augustine reflected on his own life before his conversion to Christianity, "I knew nothing about love for [God], of whom I had no conception other than of physical objects luminous with light" (Augustine, Confessions 53-4). The bishop felt that before he knew spirit directly, he knew it indirectly through its expression in the natural world, notably for the reader, in the light he saw. The consistency of the natural good he experienced in the physical world was always working on him, directing him toward the spiritual truths he would eventually accept.

> Physicists – and most everyone else as well – rely crucially upon the stability of the universe: The laws that are true today were true yesterday and will still be true tomorrow (even if we have not been clever enough to have figured them all out)... This does not mean that the universe is static; the universe certainly changes in innumerable ways from each moment to the next. Rather, it means that the laws governing such evolution are fixed and unchanging. (Greene 168)

The stability of our world is what allows us to make systematic inquiry into the nature of reality. Elsewise, reality would be a moving target, less real in each passing moment than it was in yesterday's present. Only by aggregating all the facts of past presents are we able to say with confidence if a thing must be so now or in the future. By these means we establish all physical, philosophical, and theological law. This amalgamation of miniscule moments and bits of fact are important but tell us little of the universe's realities.

> It will perhaps be agreed that if a man does wish to become master of an art or science he must go to the universal, and come to know it as well as possible; for, as we have said, it is with this that the sciences are concerned. (Aristotle 201)

Man is inclined toward the universal. We are not content to very scientifically state, 'At this one moment, with certain particular known conditions and many unknown, when this one thing acted upon this other thing in this exact way, we perceived this effect'. Professional scientists do exactly that every day, but we all would like science, theology, and philosophy to do away with this careful language and define laws that we can apply to our own lives generally. We are not content with anecdote; we crave the universal.

Each data point directs us toward the universal but cannot possibly encapsulate it all. "Each of the sentences I write is trying to say the whole thing, i.e. the same thing over and over again; it is as though they were all simply views of one object seen from different angles" (Wittgenstein, Culture and Value 7e). For the philosophical writer of any discipline who wishes to say something of worth, he must overstep the scientific boundaries of careful representation of particular data and delve into theory.

The whole, the universal, is inescapable. Even if we could, we would not want to escape. Man wants to know the universe and all its relationships. And oddly, in that knowledge the knower is somehow inseparable from the rest.

> In all of the Oriental religions great value is placed on the Sanskrit doctrine of *Tat tvam asi*, "Thou are that," which asserts that everything you think you are and everything you think you perceive are undivided. To realize fully this lack of division is to become enlightened. (Pirsig 177)

Enlightenment. Now there's a concept. To be overwhelmed by the divine light by which everything becomes clearly interrelated. To see that which has yet been unseen. To have insight into that which has been waiting to be found and to be incorporated into it all. "For nothing is hidden except to be made manifest; nor is anything secret except to come to light" (Mark 4:22). Here also, from the Vedas, ancient Hindu sacred texts:

> We have o'erpast the limit of this darkness; Dawn breaking forth
> again brings clear perception.
> She like a flatterer smiles in light for glory, and fair of face hath
> wakened to rejoice us...
> Never transgressing the divine commandments, she is beheld
> visible with the sunbeams. (Rig 1:92.6,12)

With this reference to the Hindu scriptures, it may be said that to ignore any of the major theologies and philosophies of the Near and Far East would be an egregious error when attempting to pull together a representation of all human understanding concerning the universal topic of light. We must consider the texts of Islam, Hinduism, Buddhism, Judaism, and Christianity if we are to gain elements of all knowledge that will aid us in such an undertaking, which we have indeed embarked upon. It can be said by adherents of any given philosophy or theology that his worldview is complete in itself, which may or may not be true, but in reconciling the thought of East and West, we should be surer of our conclusions.

> If someone else were to produce a thesis which purported to be a major breakthrough between Eastern and Western philosophy, between religious mysticism and scientific positivism, [we might] think it of major historical importance. (Pirsig 443)

If modern science could be reconciled into the amalgamation of eastern and western philosophies and religions, we might feel very sure of our final thesis. The reader must determine the legitimacy of this particular treatment, but it will attempt to break down the geographic and religious lines that have divided humanity for far too long. It will incorporate science into it all.

In a sense, this book will work to maintain a neutral perspective in trying to delineate what we might know about reality; however, we will soon find that one perspective offers what might be considered a comprehensive, cohesive, and compelling theory, one that in itself unites the world's major modes of thought by validating each in part and all in whole.

It should be no surprise that the means by which this unification might be done will be light itself; the energy that pervades the entire universe is passing through us all in one wavelength or another every moment, somehow composing all physical material that we encounter, uniting us in a correlation of action and enigmatic contingency.

Before we delve into, chapter by chapter, our scientific, philosophical, and theological understandings of light, let us briefly set a useful definition in place, given by the Islamic philosopher Suhrawardi:

> If you wish to have a rule regarding light, let it be that light is that which is evident in its own reality and by essence makes another evident. (Suhrawardi 81)

Each time we employ the term 'light', it will be impossible to express the totality of its meaning. In the whole of this book, the very surface of such a meaning will only be scratched, only a small corner of the impenetrable veil lifted. So, when we use the word, its comprehensive meaning must be taken for granted in all its complexity. Only after reading this text and looking into many others would one begin to appreciate all the subtle connotations and contexts that contribute to the word. But, in order to begin conversation, a certain level of empiricism will allow us to standardize different portions of the meaning that we intend at individual moments. With that thought in mind, we will begin by considering light's physical context.

MARVELOUS LIGHT

SCIENCE OF LIGHT

In the two quotes and brief timeline directly below, we can only begin to appreciate the scientific confusion concerning light that persisted for centuries, and admittedly, persists still today.

No period between remote antiquity and the end of the seventeenth century exhibited a single generally accepted view about the nature of light. Instead there were a number of competing schools and subschools, most of them espousing one variant or another of Epicurean, Aristotelian, or Platonic theory. One group took light to be particles emanating from material bodies; for another it was a modification of the medium that intervened between the body and the eye; still another explained light in terms of an interaction of the medium with an emanation from the eye; and there were other combinations and modifications besides. (Kuhn 12-13)

- 1690 - Christian Huygens
  - *Treatise on Light*
  - Light is a wave in aether, because...
  - Beams of light intersect like water waves
- 1704 - Isaac Newton
  - *Opticks*
  - Light is streams of particles, because...
  - Light shadows are unlike water wave shadows
- 1801 - Thomas Young
  - Double Slit Experiment
  - Light is a wave, because...
  - Interference is a characteristic of waves
- 1821 - Augustin-Jean Fresnel
  - Fresnel-Arago Laws
  - Light is a transverse wave, because...
  - Light can be polarized

- 1862 - James Clerk Maxwell
  - Maxwell Equations
  - Light is an electro-magnetic wave, because...
  - Maxwell Equations predict measured speed of light
- 1887 - Albert Michelson and Edward Morley
  - Michelson-Morley Experiment
  - Aether is immaterial, because...
  - Light is unaffected by earth's motion through space
- 1900 - Max Planck
  - Quantum Mechanics – E=hv
  - Light is not a continuous wave, because...
  - Atoms emit light in discrete chunks
- 1905 - Albert Einstein
  - Photoelectric Effect
  - Light is made of photons, because...
  - Light dislodges electrons in metal at certain energies
- 1905 - Albert Einstein
  - Special Theory of Relativity
  - Aether does not exist, because...
  - Relativity denies aether and is precisely confirmed
- 1913 - Niels Bohr
  - Bohr Model of Hydrogen Atom
  - Physical atom necessitates quanta, because...
  - Hydrogen emits light in predictable spectrum
- 1922 - Arthur Compton
  - Compton Effect
  - Photons are physical, because...
  - X-rays bounce off atoms like physical collisions
- 1924 - Louis de Broglie
  - $\lambda = h/p$
  - All elementary particles are wave-like, because...
  - Equation accurately relates mass to wavelength

Since the dawn of modern science in the sixteenth and seventeenth centuries, light has been pictured either as particles or waves – incompatible models – each of which enjoyed a period of prominence among the scientific community. In the twentieth century it became clear that somehow light was both wave and particle, yet it was precisely neither. For some time, this perplexing state of affairs, referred to as the *wave-particle duality*, motivated the greatest scientific minds of our age to find a resolution to apparently contradictory models of light. (Pedrotti and Pedrotti 1)

We will revisit many of the facts in the above quotes and timeline in great detail in the pages to come as well as much more that would not fit easily into outline form, so please do not get hung up on anything that might be confusing you now. Detail is coming.

Largely due to the way our scientific knowledge is learned from elementary school up through undergrad, students of science often assume that the totality of our current scientific understanding is organized into one systematic whole. The way textbooks are presented would suggest this, though it is not the case.

We so readily assume that discovering, like seeing or touching, should be unequivocally attributable to an individual and to a moment in time. But the latter attribution is always impossible, and the former often is as well. (Kuhn 55)

Scientific discovery is active and dynamic, as is our scientific understanding. It is not so easy to parse things up into digestible bits. We do discredit to scientists themselves to assume one man at one time discovers profound truths. Science is a fluid process contributed to by countless dedicated individuals over time.

Scientific discoveries reassert facts and predictions that become probable, but scientific confirmation takes a long time to establish what we consider 'law'. In spite of all of the experimental confirmations that light is a wave, James Clerk Maxwell contributed to the affirmation of the concept in his proposal of mathematical equations that matched the measured speed of light when he assumed that light is an electromagnetic wave. Likewise, Einstein himself was still having to make comment on the 'luminous aether' twenty years after Michelson and Morley demonstrated that the aether was immaterial. The modern reader might be surprised that such an ancient concept had to be refuted in 1905. That fact is true because scientific progress is slow, even when scientific discovery can occur in fits and starts.

Even Einstein, due to the immense cultural and professional respect he gained through his work, was able to stymie general acceptance of some difficult concepts in quantum mechanics. Einstein believed that 'God doesn't play dice with the world' but went on to accept the proofs in which Heisenberg, Bohr, and Schrödinger suggested that probabilities and uncertainties were intrinsic to the basic fabric of our physical universe. Einstein's conceptions of the 'Old One', his belief in the divine, and his philosophical predilections kept him, much of the scientific community, and large swaths of general culture from believing what only slowly became inescapable physical fact. However,

> By 1927, therefore, classical innocence had been lost. Gone were the days of a clockwork universe whose individual constituents were set in motion at some moment in the past and obediently fulfilled their inescapable, uniquely determined destiny. According to quantum mechanics, the universe evolves according to a rigorous and precise mathematical formalism, but this framework determines only the probability that any particular functions will happen – not which future actually ensues. (Greene 107)

This was an uncomfortable step away from determinism for Newtonian scientists, as well as many modern philosophers and theologians who cling devoutly to deterministic concepts like unconquerable fate and predestination. But the modern application of probabilities must be considered; we must contend with them because of their remarkable ability to predict and explain what we consistently observe in our world. Like Einstein, let us have the humility to accept that we have not yet attained perfect knowledge:

> Reluctantly, Einstein conceded the technical correctness of the system Heisenberg and Bohr laid out. But he could never accept that it was the last word... Heisenberg's uncertainty, Einstein stoutly insisted, was a sign of human inability to comprehend the physical world, not an indication of something strange and inaccessible about the world itself. (Lindley 6)

Today we have come to a point where we must consider what earned Schrödinger the 1933 Nobel Prize in Physics: physical substance may only be a superficial manifestation of waves and fields, perhaps as string-theory suggests, of one-dimensional vibrating strings that produce those energies. In other words, matter is simply another expression of energy. Einstein resisted acceptance of the thought, but modern science tells us that our physical world is not as tangible and as certain as we have always assumed. Light is the key that allowed us to peer into this mystery and to witness what lies behind the physical.

WHAT IS LIGHT?

Let us begin with the most basic aspects of light. Where does it come from?

Every single physical thing in the entire universe emits light. So long as a molecule exists, it is active. So long as a molecule is active, it will release energies along the electromagnetic spectrum, e.g., infrared radiation (IR):

The molecules of any object at a temperature above absolute zero (-273°C) will radiate IR, even if only weakly. On the other hand, infrared is copiously emitted in a continuous spectrum from hot bodies… A common light bulb actually radiates far more IR than [visible] light. (Hecht 70-71)

Although we associate sunlight with the illumination it provides the world, the Sun actually gives off more than half its light in infrared. This isn't terribly surprising when you think about how we also feel heat from that same sunlight, which is, in fact, due to the infrared light being delivered from the Sun. (Arcand and Watzke 66)

Because everything around us – the air we breathe, even the materials we use to build with, are composed of spinning and vibrating atomic particles, you and I are literally swimming in a turbulent sea of electromagnetic fields. We are part of it. (Taylor 20)

Light, in one form or another, is all around us at all times. It is in us and flowing through us and being emitted by us. Light is ubiquitous. It is inescapable. Everything, regardless of its temperature, is producing light in some wavelength along the spectrum of electromagnetism.

The wavelength of light emitted often depends very closely on the temperature of the object that is emitting it. For this reason, it's often useful to visualize the spectrum of electromagnetic radiation as a thermometer. (Arcand and Watzke 10)

From the extremely long wavelengths of radio waves to the highest frequencies of gamma rays, light's wavelength is an attribute closely linked to the temperature of that which released the wave. Only very hot and/or radioactive objects can emit gamma rays, while most materials might emit radio waves. A helpful way to visualize what we are talking about is by narrowing the breadth of electromagnetic frequencies to only visible light, measured in terahertz (THz). Consider visible light waves and their effect on the human physiology:

Colors are the subjective human physiological and psychological responses, primarily, to the various frequency regions extending from about 384 THz for red, through orange, yellow, green and blue, to violet at about 769 THz. Color is not a property of the light itself but a manifestation of the electrochemical sensing system – eye, nerves, brain. (Hecht 72)

Particular wavelengths of electromagnetic radiation are emitted by different objects, and those waves interact with the human eye and other physiological systems in different ways depending on the wavelength. But, no matter the wavelength or color, whether radio waves or microwaves, infrared or visible or ultraviolet light, x-rays or gamma rays; no matter its effect on atoms or molecules or the human physiology and psychology; "no matter their dissimilarities, light is light is light" (Arcand and Watzke 23). All electromagnetic radiation is the same physical phenomenon, simply at different wavelengths.

Naturally, different wavelengths will have different effects on atoms just as they have different effects on all matter, depending on the properties of the material.

An atom can react to incoming light in two different ways, depending on the incident frequency or equivalently on the incoming photon energy ($E=hv$). Generally, the atom will "scatter" the light, redirecting it without otherwise altering it. On the other hand, if the photon's energy matches that of one of the excited states, the atom will "absorb" the light, making a quantum jump to that higher energy level. (Hecht 57)

When the electron falls back to its original orbit, the energy that gets released is a packet of light known as a photon. When these photons are created, they come out of the atomic womb blazing at the speed of light (about 186,000 miles per second). (Arcand and Watzke 14)

The excited state referred to above is the excited state of the electrons found in each atom. The electrons, orbiting the nucleus at incredible speeds, form a cloud around the tightly packed protons and neutrons and offer the photon an entity with which to interact. Incoming photons of light may simply be redirected by the nucleus; however, they may also share their energies with the electrons, affecting the balance of the atom.

Once the electron cloud starts to vibrate with respect to the positive nucleus, the system constitutes an oscillating dipole and so will presumably immediately begin to radiate at that same frequency. The resulting scattered light consists of a photon that sails off in some direction carrying the same amount of energy as did the incident photon – the scattering is elastic. (Hecht 57)

This 'oscillating dipole' introduces the primary features of electromagnetic radiation. The nucleus of an atom carries a positive charge due to the protons found in it. Conversely, the electron cloud carries a negative charge due to the electrons themselves. The balance between the opposite charges of nucleus and cloud constitutes a dipole. When the electrons shake and vibrate in relation to the protons, they are said to oscillate. The vibrating charges create a varying electric field.

Scientifically, we know that variations in the position of electronic charges create magnetic fields. Likewise, variations in magnetic fields create electric fields, each influencing the other back and forth, on and on. Without delving into the calculus, the changes in the fields (the derivatives) are not constant in time, and this inconsistency is what propagates the electromagnetic relationship into perpetuity.

At the instant the charge begins to move, the E-field [electronic field] is altered in the vicinity of the charge, and this alteration propagates out into space at some finite speed. The time-varying electric field induces a magnetic field. But the charge is accelerating, dE/dt is itself not constant, so the induced B-field [magnetic field] is time dependent. The time-varying B-field generates an E-field, and the process continues, with E and B coupled in the form of a pulse. As one field changes, it generates a new field that extends a bit further, and the pulse moves out from one point to the next through space... The E- and B-fields can more appropriately be considered as two aspects of a single physical phenomenon, the *electromagnetic field*, whose source is a moving charge. (Hecht 40)

The electromagnetic field is light itself, expressed differently at different wavelengths, but always traveling 'at some finite speed', the limit of which is the speed at which fields extend in a vacuum, the speed of light, 299,792,458 meters per second. The E and B fields constantly regenerate each other at the speed of light, continuing the process infinitely through space and time until the pulse encounters packets of matter that absorb, redirect, or slow the electromagnetic energy.

THE QUANTUM ENIGMA

In just a few pages, we have given a fair treatment to the very basics of light. Though they are easy to overlook, we have already encountered a few of the enigmas that have haunted physicists through the past century and that we will be focused on in much of the rest of the current text.

Light is an electromagnetic pulse, a wave of energy, that travels through the vacuum of space, always at a set speed. Light is a wave. But not exactly in the ways we understand other waves.

A water wave in profile and a wave in the battle ropes at your CrossFit gym are transverse waves, moving up and down in succession. A sound wave is a longitudinal wave, expanding and compressing molecules of air into pockets of high and low density that push forward in succession. Transverse and longitudinal waves have some similar properties, some dissimilar, but the primary way that we understand these waves is how they physically move through their respective mediums: air, water, rope, etc.

Electromagnetic radiation, however, is capable of moving through the vacuum of space. This should strike us as odd. It did the same to the ancients. The mysterious propagation of light is the physical oddity that was the source of the concept of the luminous aether. Man has observed the wave-like properties of light for centuries, and so we understood it as a wave. But every other wave we knew was only a wave insofar as the medium it moved through composed the wave. We could not comprehend a wave that had no medium, so we theorized about an unseen, immeasurable medium: aether. It turns out that we could neither see nor measure aether because it did not exist.

Yet, as we did away with the problematic theory of the luminous aether, we could not do away with the problems that remained without it. Light is a wave without a medium. This is very strange.

We may be able to accept that light is a wave without the need of a physical medium, but we know that it is also a photon, a physical packet of energy. Newton believed light was a stream of physical particles moving through the air. Though many experiments between his time and the early 1900s confirmed that light is a wave, Einstein reintroduced Newton's concept of light as a particle when he described the photoelectric effect.

When light is directed at and through thin metal foils, the valence (free floating) electrons of the metal are dislodged and scattered. In understanding light as a wave, as the amplitude (or intensity) of the wave is increased, dislodged electrons should fly away with more energy transferred from the higher energy waves. But this is not what happens.

Instead, as the intensity increases, the wave dislodges more electrons with the same energy. If we want to dislodge electrons of higher energy, we must shoot a higher frequency of electromagnetic radiation at the metal. This behavior would not result from a wave. This is the photoelectric effect.

Einstein reconciled light and the photoelectric effect by thinking about light as a physical thing, a photon. In his model, by increasing the intensity of the light, we are simply shooting more photons of the same energy at the foil, dislodging more electrons of the same energy, which is what we observe. If we want to transfer more energy into the electrons, we must use photons of a higher frequency, the true measurement of the energy of light. The intensity of the light, the amplitude of the wave, and the number of photons are the same variable in our experience of light. On the other hand, the wavelength ('color') of light, the frequency of the wave, and the energy/momentum of the photon are another variable.

In order to understand the photoelectric effect, an observed scientific fact, we must understand light as a photon. In order to understand many other characteristics of light, like those employed in optics, we must understand light as a wave. Depending on the effect we are measuring, we have to measure light by different means. This is an enigma and a paradox.

Modern physics has encountered some of the greatest paradoxes the world has ever known. Experimentally we have verified characteristics of light that have always been assumed to be mutually exclusive and contradictory; however, light displays both properties, so we must work to reconcile the experimental results. The experiments themselves sometimes seem untrustworthy.

> Few philosophers of science still seek absolute criteria for the verification of scientific theories. Noting that no theory can ever be exposed to all possible relevant tests, they ask not whether a theory has been verified but rather about its probability in the light of the evidence that actually exists. (Kuhn 144)

Scientists today do not believe that they will ever prove a theory as a law beyond any doubt. There always remains against every theory the possibility of a phenomenon that would falsify the interpretation of past results. However, we believe that the universe is ruled by physical laws, even if we do not know those laws. We believe that those laws abide.

> The character of the belief in the uniformity of nature can perhaps be seen most clearly in the case in which we fear what we expect. Nothing could induce me to put my hand into a flame – although after all it is *only in the past* that I have burnt myself. (Wittgenstein, Philosophical Investigations 134e)

The careful philosopher of science would not say beyond any conceivable bit of doubt that you will definitely be burned by flame. There is still the theoretical possibility that it may not burn you based on some physical law or variable that is unknown to us. However, we do not put our hand into the fire because theoretical possibilities lose their luster in the face of constant and unquestioned uniformity of experience. In every other situation we've ever known, the fire burned, so we correctly judge that it will burn again.

In similar ways to how we cannot *theoretically confirm* that a particular thing will happen, but we *practically know* that it will, we also believe in the mathematical and theoretical constructs of the scientist because reason tells us that they are logical, coherent, and sufficiently predictive.

We do not experience the atom directly and thus cannot describe it perfectly. Any attempt to do so with current observation techniques will not be quite scientific, no matter how earnestly we try. Niels Bohr, who developed what is now accepted as the atomic model of the atom, said in describing the model:

> When it comes to atoms, language can be used only as in poetry. The poet, too, is not nearly so concerned with describing facts as with creating images and establishing mental connections. (Lindley 86)

Bohr is saying that any picture drawn of a physical construct might not and should not be expected to describe perfectly the experimentally collected data. The model is only a picture. In this vein of thinking, an old engineering quip humorously states, "All models are wrong. Some models are useful."

So, when we encounter the enigmas of light and quantum dynamics, we should not be concerned that we cannot obtain a perfect model of the enigma. We cannot picture what a physical thing like a photon would look like passing through space like a wave at a speed that no other physical thing can reach. There is no model for this enigma, but even if there was, what we believe is a predictive mathematical formula as opposed to a poetic description of the formula. In light of the practical usefulness of the mathematical model, we grow less concerned that:

> Classical electrodynamics, as we shall see, unalterably leads to the picture of a continuous transfer of energy by way of electromagnetic waves. In contrast, the more modern view of quantum electrodynamics describes electromagnetic interactions and the transport of energy in terms of massless elementary "particles" known as *photons*, which are localized quanta of energy. (Hecht 33)

It may take more significant effort for some readers to come to accept this thought than others. We must think of physical light in two ways, for no one way encapsulates all the observed characteristics. It is very likely that in the current text there will not be enough information for the doubter to accept this enigma. All I can do is to say that short of running these classical quantum experiments for oneself, which would require much education, time, and expense, each person will have to defer to an authority on which he or she will place trust. It will be easiest to accept the testimony of exceptionally intelligent physicists than to live in endless skepticism. Using the theories and equations they have developed, we have established an unquestionably valid basis of confirmed experimental outcomes.

> Beyond the fact that it is a mathematically coherent theory, the only reason we believe in quantum mechanics is because it yields predictions that have been verified to astounding accuracy. (Greene 88)

This *quantum* world we refer to is the one in which atoms are only able to absorb and emit light in particular spectra. Classical wave dynamics do not permit this world to exist, yet the results of quantum experimentation confirm its existence. Again, Bohr, who contributed significantly to the quantum theory, is believed to be the one who said, "There is no quantum world. There is only an abstract quantum description. It is wrong to think that the task of physics is to find out how nature *is*. Physics concerns what we can *say* about nature" (Rosenblum and Kuttner 131). And in this case, what we can say about nature is the results of its behavior; its mysterious, enigmatic, quantum behavior.

Heisenberg's uncertainty principle goes one step further in restricting what can be known and said about the quantum world, the realm of photons. The uncertainty principle proved that "measuring one aspect of a [quantum] system closes the door on what else you can find out, and thus fatally restricts the information that any future measurement might yield" (Lindley 155). In other words, the act of observing the quantum world changes the thing observed so that no other observation of another feature of the thing at the same time will be valid. Bohr and Heisenberg worked together within these means to develop the criticized, yet generally accepted Copenhagen Interpretation of quantum mechanics, which, "argues that since we never deal *directly* with the quantum objects of the microscopic realm, we need not worry about their physical reality, or lack of it" (Rosenblum and Kuttner 126). This interpretation encapsulates the complementarity principle of Niels Bohr,

> According to which both the wave aspect and the particle aspect of quantum objects had necessary but contradictory roles to play. Depending on the problem, one aspect or another might come to the fore, but neither could be neglected entirely. (Lindley 148)

If quantum photons and the enigmas of light had been the end of the issues in this new quantum world, light would have stood alone as a difficulty; however, the developments that came about because of the research on light complicated the very roots of all physicality in our universe. The trouble continues with Max Planck and his constant, h.

> Planck found that by adjusting *one* parameter that entered into his new calculations, he could predict accurately the measured energy of [a closed system] for any selected temperature. This one parameter is the proportionality factor between the frequency of a wave and the minimal lump of energy it can have. (Greene 93)

Planck's constant is the conversion factor which gives a mathematical basis for the observed quantum 'jumps' that contradict the expected classical continuity of waves. The atom refuses to be treated with continuities, and Planck's constant turned out to be the conversion factor between frequency and energy in the photon. The work of Louis de Broglie complicated matters further when he employed Planck's constant in his equation, $\lambda=h/p$, to determine the relationship between an elementary particle's wavelength and its momentum. Both momentum and wavelength are measurements of energy, but wavelength measures a wave's energy and momentum measures the energy of physical matter. Planck's constant related the two and established the relationship between wave and particle that could then be used to mathematically explain that all elementary particles, like electrons and protons, are similar to photons in their enigmatic qualities. The non-sensical difficulties that had once only applied to light now apply to everything that composes our physical universe. In our understanding of light, we have somehow unlocked the physical basis of everything. That fact is very unsettling, though it encourages science onward.

With every page turned, the complexities of understanding the nature of light multiply to the point that we now have to extend our discussion to encapsulate all physicality. The minutest realms we have peered into contain the very building blocks of the totality of our physical world, and what we have been saying about those blocks is terribly difficult. We have the means to describe the micro-universe in such a way that allows us to develop equations that predict experimental results with uncompromising and unparalleled accuracy. Using quantum physics and Einsteinian relativity (another enigma treated later), scientists have theorized and confirmed many new features of our world. But still, the theories are unsatisfying. We use the equations and extract the data, but we do not quite believe what they suggest about nature. "We have confronted the still unresolved, and definitely controversial, quantum enigma. However, the *experimental* results we have described are completely undisputed" (Rosenblum and Kuttner 99).

It seems that we will not be able to escape the logical impossibilities of the quantum world. Each side of the paradox is sufficiently well-established and mathematically defensible. If we cannot rid ourselves of the mysteries illuminated by light, we will have to join in and do our best to apply the consequences to our own worldview.

SCIENCE IN CRISIS

It is an interesting, intuitive, and unlikable (from the inside) quality of scientific paradigm shifts that allows for 'outsiders' to enter an area of study and destabilize the established order by conceptualizing and theorizing new and better ideas by which the science may progress. These agents of chaos are usually inexperienced in the field, not having received a wholly typical education by which they would have learned to toe the paradigm's line. Historically, they are relatively young men with the imaginative ability to see things in ways they had not been seen before. The early 1900s were riddled with these men upsetting the established order, upsetting centuries of assumption that presumed a deterministic universe, upsetting each other, and turning the world of physics on its head. From the man who literally wrote the book on paradigm shift, Thomas Kuhn:

> Almost always the men who achieve these fundamental inventions of a new paradigm have been either very young or very new to the field whose paradigm they change. And perhaps that point need not have been made explicit, for obviously these are the men who, being little committed by prior practice to the traditional rules of normal science, are particularly likely to see that those rules no longer define a playable game and to conceive another set that can replace them. (Kuhn 90)

Einstein, the man whose name now universally represents genius, worked in a patent office as a clerk while he drafted his special theory of relativity. Of course, no one of the established academic order could have predicted this completely unknown man's meteoric rise into the scientific and cultural limelight. Nor could anyone have predicted the consequences.

As if the quantum enigma did not cause enough strife in the volatile state of physics in the twentieth century, Einstein's work threw another significant wrench into the machinery.

As they are currently formulated, general relativity and quantum mechanics *cannot both be right*. The two theories underlying the tremendous progress of physics during the last hundred years – progress that has explained the expansion of the heavens and the fundamental structure of matter – are mutually incompatible. (Greene 3)

Mutually incompatible. Based on the well-established study of logic and the theoretical corollaries of the two models, quantum mechanics and relativity cannot describe the same universe. The premises of the one exclude those of the other. They cannot both be right.

Unfortunately, based on current understanding of scientific verifiability, both theories are indeed right. We keep coming back to the question 'How can that be?' We have two theories with unrivaled predictive power that are confirmed time and time again by experimental results. And yet the theoretical ends of the two are incompatible. How?

That question haunted Einstein's contemporaries and still haunts all honest scientists today. String theory is working to incorporate the two theories into one, but the hope of its success is theoretical, inconclusive, and easy to doubt. We are so unsure of what lies ahead. In 1910, Planck reported that physicists "now work with an audacity unheard of in earlier times; at present no physical law is considered assured beyond doubt, each and every physical law is open to dispute" (Lindley 67). Even though quantum theory and relativity have been established individually in modern science, the existential discomfort between the two lives on.

In their consternation, physicists might desperately seek some wisdom from philosophy. They may be ready to hear Socrates' opinion: "While one judgement cannot be *truer* than another, it can be *better*, in the sense of having better consequences. This suggests pragmatism" (Russell 151). Pragmatism allows us to stop making big, concrete truth claims and instead focus on what is helpful as opposed to that which is unequivocally correct. But unfortunately, raw pragmatism is no help in modern physics' current predicament. Quantum theory and relativity have equally well-confirmed consequences. Both are 'better' than the other at times, regardless of which is 'truer'. It is only in their comparison that the joint-consequences sour.

If we adopt a purely utilitarian view and make no demands on the quantum theory beyond the successful prediction of experimental results, the discussion could end here. But man undertakes the study of nature not merely to master it, but out of a deep-seated need to clothe his life experience in meaning. (March 239-240)

It is difficult to shake the above sentiment. Science has undoubtedly improved modern life through health, convenience, and productivity; however, man wants answers that science might not be able to provide. There are some who can overcome this urge:

The task of science [is] solely to find connections between measurements. To inquire into the reality that lay behind these connections, to ponder such hazy concepts as cause and effect, [is] regarded as both futile and unscientific, possibly no more than an exercise in language. (March 222)

"It is wrong to think that the task of physicists is to find out how nature is. Physics concerns what we can say about nature." (Lindley 196)
-Niels Bohr

"Whereof we cannot speak, thereof we must be silent." (Lindley 196)
-Ludwig Wittgenstein

Unfortunately, we refuse to stay silent. As mentioned before, man has an insatiable need to understand the troublingly incomprehensible reality in which we live. Even in pragmatic studies like physics, some sort of 'ultimate' question continues to arise time and time again, a question we cannot answer.

So we reach an impasse. Classical physics cannot say why the universe happened, because nothing can happen except that prior events caused it to happen. Quantum physics cannot say why the universe happened, except to say that it just did, spontaneously, as a matter of probability rather than certainty. (Lindley 219)

If the modern reader allows herself to disregard Einstein for a moment, it may seem that we should rid ourselves of the classical formulations of these problems. Newtonian physics is a helpful tool, but perhaps it should be tossed aside so that quantum physics, the more real explanation of our world, can stand alone as the universal theory. But we cannot do that.

> Though revolutionary in conception, the end result of Einstein's relativity was to save classical physics by reformulating it in a manner consistent with the known properties of light. (March 158)

Without discussing much about relativity in this chapter, March's quote allows us to appreciate our problem. Quantum physics toppled classical physics as a limited understanding of our physical world; however, relativity simultaneously pulls Newton out of the ashes by restructuring the physical concepts of space, time, energy, and matter into a far more comprehensive theory. Einstein's theory is correct, as far as modern science allows us to test, but it does not solve the issue. It complicates modern science even further. As if it was not complicated enough already.

SCHRÖDINGER'S CAT

This book will not expound upon a complete scientific treatment of the wave-particle duality of light, but it is important to understand the results of the duality, which will serve as an acceptable introduction to the concept.

Insofar as light is a wave, and as long as we do not attempt to locate its associated photon, the *wave function* (a mathematical construct of the 'waviness' of a photon, though difficult to imagine physically) allows it to delocalize across a range of physical locations. The wave function is the feature of the duality that permits light to interact with itself in the double slit experiment and cause an interference pattern on a screen on the other side of the barrier bearing the slits. After passing through the slits, light strikes and illuminates the screen in a predictable pattern of bright and dark bands, different depending on the light's wavelength, but always confirming the wave-nature of light.

The peculiar thing about the double slit experiment is not necessarily that light interferes with itself. Sound waves do the same thing when emanating from two sources, and we can understand why. The peak of one wave is 'cancelled out' by the trough of another, and the combination of two peaks or two troughs increase the amplitude of the interacting waves. There are locations around two speakers with greater amplitude than either single sound wave, and locations where the cancelled waves are flat. In a room with two speakers playing the same thing, there will be areas of 'cancelled' silence and areas louder than one speaker is on its own. We can conceptualize two sound waves interacting in this way without much difficulty.

What is so difficult to understand about light is that a single photon will interfere with itself. Of course, after passing through the slits of the double slit experiment, one photon will only strike one location on the screen, and one instance cannot express an interference pattern of 'peaks' and 'valleys', but the generalized results of many single-photon events spread out into an interference pattern as if a stream of photons were sent through the slits at once, interfering with each other. In some locations, many photons strike the screen; in others, none. But the individual photons did not have the opportunity to interfere with any other light. Each photon interfered with itself, one at a time, creating the signature pattern. That is very odd.

Leaving the wave-nature behind for a moment, insofar as light is a photon, we can be sure of its location. If given two paths to travel in the necessary apparatus, we could determine which of the two paths the photon traversed, as a discrete physical thing could only be at one place at one time. But having determined that the photon travelled a particular path, we would collapse the wave function and disrupt the interference pattern. When we know which slit the photon passes through, there is no interference at all and the photons strike the screen at random. We do not give the wave function the opportunity to interfere with itself between the two slits if we know it only passed through one slit. If we know the location of the photon, it ceases to behave like a wave.

If we measure light as a wave, it behaves like a wave. If we measure light as a photon, it behaves like a photon.

Schrödinger's Cat is a famous thought experiment to be set up as follows: a cat is placed in a box. That box has two openings on one side forming a double slit experiment. In Slit 1 is a device that will detect the photon if it passes through that slit. Connected to that device is a mechanism that will break a vial of poison if the detection device is triggered. If the photon passes through Slit 1, the vial will be broken and the cat will die. If the photon passes through Slit 2, nothing will happen and the cat will live.

The wave-particle duality of light necessitates, however, that an observer observes which slit the photon passes through in order for the photon to actually pass through one or the other slit. Until that is observed, the wave function of the photon is split between the two slits, and the photon has a 50/50 probability of yet passing through either slit. Even after the photon has entered the box, we cannot say which slit it passed through until we open the box to find out if the cat is dead. If we do not open the box, the cat remains in a *superimposed* state in which it is both dead and alive, contingent on the indeterminate 'waviness' of light. This is not the same thing as saying there is a 50/50 chance that the cat is dead or alive. In the superimposed state, the cat is both fully dead and fully alive, not yet what it already is.

The system awaits our observation to determine which state was always the result. The past is not determined until a future observation necessitates that it has been determined. Our observation of the system, and thus our observation of the path the photon took, is necessary for the photon to actually take the path. Until we observe the result, the photon's wave function is spread out between the slits as if it passed through neither and both.

Because Schrödinger's Cat is a thought experiment, the physicist can rid the experiment of the complications of reality. In reality, all physical entities and their wave functions are entangled in such a way that the cat's body, dead or alive, interacts with its box and its box interacts with the outside world. The results of the cat's life or death inside the box is expressed imperceptibly in the world outside the box regardless of whether a sentient being notices the expressions or not. In interacting with the rest of the world, the cat has thus been 'observed'. However, in the instant light passes through the slits, the light is not yet interacting physically with anything else, and the experiment is valid.

Beyond that single instant, the full validity of Schrödinger's Cat depends on a system that is completely separable from the rest of the universe. The concept of separability has since been determined to be impossible due to the utter entanglement of all particles in the universe via light. This complicates the implications of Schrödinger's experiment beyond human comprehension.

INESCAPABLE ENIGMAS

Probability has not disappeared, Schrödinger's cat still has a fifty-fifty chance of being found alive when the box is opened. Beyond that, nothing more can be said. That, ultimately, is what so distressed Einstein – the idea that physical outcomes are truly unpredictable. Physicists today who share that distress cannot shake the feeling that something must be missing, that quantum mechanics must be, as Einstein [et. al.] said, incomplete. On the other hand, no experiment has yet found a flaw in quantum mechanics, and no theorist has come up with a better theory. (Lindley 199)

"Quantum physics is very imposing. But an inner voice tells me that it is not the real McCoy. The theory delivers a lot but hardly brings us closer to the secret of the Old One. I for one am convinced that *He* does not throw dice." (Lindley 137)
-Albert Einstein

A connection awaited discovery in the beginning of the nineteenth century, and we still wait for a real connection today. As mentioned, string theory is a somewhat promising theory at present, but it has yet to catch up with the raw necessity of quantum physics and relativity as predictive theories. Until the connection is found, we simply have to keep pushing ahead with the impossibilities of quantum mechanics and relativity as individual theories, as well as the difficulties of living in a world where both are simultaneously valid.

"Neither of the two concepts must be discarded, they must be amalgamated. Which aspect obtrudes itself depends not on the physical object, but on the experimental device set up to examine it." (Hecht 33)
-Erwin Schrödinger

Schrödinger was right. We live in a world in which two mutually contradictory theories are both completely valid and correct and useful. We live in a world in which those individual theories are wholly contradictory in themselves. We live in a world in which, depending on what it is we are studying or experimenting, we will employ one theory or another to test what it is we want to test, limiting our ability to test anything else. No one voice is allowed to speak over another. We must have all the contradictions, for only through them do we have a complete view of this very confusing world.

It is helpful for the modern audience to parse the world into three realms to understand the distinction between the theories that we most often employ today. Firstly, we have our commonplace, everyday world. This is Newton's domain. Newton's theories have been shown to be far from complete, but in the everyday world, his understanding of physics is so sufficiently predictive that we continue to use Newtonian equations in many ways today. Civil engineers do not worry themselves with the complexities of Einsteinian relativity when Newton will do just fine.

When we look to the stars, however, when we consider the great expanses and distances and speeds of the *cosmos*, Newton's theories fall to bits. His equations may still work fairly well within our solar system, but eventually the predictive quality of Newtonian mechanics crumbles. Here we need Einsteinian relativity to deal with great distances and the curvatures and connections of space and time. Newton will not do.

Likewise, when we peer past where our microscopes can reach, when we look to the *miniscule*, Newton cannot comment. Here we need quantum mechanics to explain the complexities of a paradoxical world. Classical physics will not do.

> Classical physics explains the world quite well; it's just the "details" it can't handle. Quantum physics handles the "details" perfectly; it's just the world it can't explain. You can see why Einstein was troubled. (Rosenblum and Kuttner 7)

> For big things (macroscopic) "their classical description is only an *approximation* to the correct quantum laws of physics. If so, the microscopic realm, the *unobserved* realm, is in some sense the more real. [Philosophers and theologians] would like that." (Rosenblum and Kuttner 131)

So, we need quantum mechanics and relativity to comment on the extremities of our physical world where Newtonian physics breaks down. However, if we try to extend those theories beyond their extreme limits, we find that when they overlap, logical and mathematical impossibilities ensue.

We have to divide our universe into bits and use different theoretical frameworks to deal with each. This is supremely unsatisfying. We are not content to be so utilitarian in our approach to the world. We want something that will encapsulate it all, however, in communicating with Einstein:

> Heisenberg insisted, it was no good imagining you could construct an absolute, God's-eye view into the inside of the atom. You could only observe in various ways the atom's behavior – the light it

absorbed and emitted – and infer as best you could what was going on inside. (Lindley 132)

Einstein had a tough time accepting that point of view. As a serious believer in the divine, Einstein felt that nature was absolute and that our knowledge of nature could be just as sure. Uncertainty and wave functions did not fit in well with Einstein's scientific philosophy. His own theory of special relativity certainly contained difficulties when considering the speed of light, but he could satisfactorily conceptualize the theoretical consequences of relativity. "The quantum enigma, on the other hand, arises directly from experiment. It's harder to ignore an enigma arising directly from experimental observation than one arising only from theory" (Rosenblum and Kuttner 97). Even so, Einstein's theory, though not developed to justify experimental observation, has been experimentally confirmed since his time. We all feel the frustration of trying to understand these theories and their consequences.

Furthermore, quantum theory and the concept of entanglement somehow also incorporate the consciousness of observers, which is a consequence never imagined to be possible in physics. The pragmatism that was necessary to accept our current state of affairs in theoretical physics does little to prepare us for the idea that the scientist observing an experiment becomes an integral part of the experiment. That interactive aspect of quantum physics escapes the pragmatic and forces us to deal with the *personal*, which we had removed from physics along with the idea of purpose. Consciousness should have no part in the apparatus of the purely physical, yet it does, and consequently, as theoretical physicist Andrei Linde asks:

> Will it not turn out, with the further development of science, that the study of the universe and the study of consciousness will be inseparably linked, and that ultimate progress in the one will be impossible without progress in the other?... will the next important step be the development of a unified approach to our entire world, including the world of consciousness? (Rosenblum and Kuttner 264)

Only after fully appreciating the theory of quantum mechanics and the wave-particle duality do we find that "it is not so easy to separate the questions of knowledge and existence. They have a maddeningly intimate connection" (March 234). As odd as this relationship is in science, consciousness and existence as complementary ideas are not at all foreign in the world of theology. In the Bhagavad Gita, Krishna comments:

> Arjuna, know that anything
> inanimate or alive with motion
> is born of the union
> of the field and its knower. (Gita 13:26)

The fact that physics, through quantum mechanics, began to go down a path on which science has less authority than the 'lesser' study of theology led some scientists to betray the determinism that has dominated their field for centuries. Many became more and more careful to comment only on what they were sure.

> As Bohr spoke more and more enigmatically on increasingly wide-ranging subjects, his determination not to say anything straightforward or concise begins to seem almost a phobia, a psychological hang-up. (Lindley 203)

Toward the end of his career, Bohr may have been becoming more careful about what he might say about the physical world, but others maintained the same determination to understand the universe as it is and to say definitive things about it. Schrödinger, whose work did so much to destabilize scientific thought, held:

> Bohr's standpoint, that a space-time description is impossible, I reject at the outset. Physics does not consist only of atomic research, science does not only consist of physics, and life does not only consist of science. The aim of atomic research is to fit our empirical knowledge concerning it into our other thinking. All of this thinking, so far as it concerns the outer world, is active in space and time. If it cannot be fitted into space and time, then it fails in its whole aim, and one does not know what purpose it really serves. (Rosenblum and Kuttner 131)

Schrödinger's perspective is extremely practical and does not share the careful pragmatism of pure scientific statement. His is a refreshing thought after chasing theoretical physics to such dizzying depths. Another practical perspective is that of John Stewart Bell, a more modern quantum physicist, who felt, "Quantum mechanics reveals the incompleteness of our worldview... it is likely 'that the new way of seeing things will involve an imaginative leap that will astonish us'" (Rosenblum and Kuttner 101).

One might be surprised to find Henry David Thoreau, an American philosopher who died in 1862, making this surprisingly modern statement:

> Now we know only a few laws, and our result is vitiated, not, of course, by any confusion or irregularity in nature, but by our ignorance of essential elements in the calculation. Our notions of law and harmony are commonly confined to those instances which we detect; but the harmony which results from a far greater number of seemingly conflicting but really concurring laws, which we have not detected, is still more wonderful. The particular laws are as our points of view, as, to the traveler, a mountain outline varies with every step, and it has an infinite number of profiles, though absolutely but one form. Even when cleft or bored through it is not comprehended in its entirety. (Thoreau 306)

In this quote we find the source of Pirsig's insights quoted previously in the introduction. The mountain of reality, even limited to physical reality, is far too expansive to contain within the limits of our current scientific understanding. We do not yet understand much concerning the physical world, and there is a certain beauty in this fact.

In his acceptance speech for the 1969 Nobel Prize in Physics, Murray Gell-Mann expressed that, "Bohr [had] brainwashed generations of physicists into believing the problem has been solved" (Rosenblum and Kuttner 138). Gell-Mann was referring to Bohr's interpretation of quantum mechanics as opposed to the uncertainty which characterized his late career. Gell-Mann was saying that there is yet so much that we do not know about the structure of the atom and how the quantum world functions, and we should not pretend otherwise.

You may have supposed after your high school chemistry class that pretty much all we could know about the atom has been neatly tucked away into encyclopedic collections of academic libraries, but that is simply not the case. Quantum, relativistic, and the combination of their enigmas are ghosts that still haunt the halls of collegiate physics departments and theoretical laboratories the world over. If we cannot yet find the answers we seek from science, let us see what the other departments of our global university have to say about light.

Let us begin with philosophy and transition in the next chapter to what some of the world's greatest thinkers have called 'The Divine Light of Reason'.

PHILOSOPHY OF LIGHT

Science may set limits to knowledge, but should not set limits to imagination. (Russell 16)

In the creation of the heavens and earth; in the alternation of night and day; in the ships that sail the seas with goods for people; in the water which God sends down from the sky to give life to the earth when it has been barren, scattering all kinds of creatures over it; in the changing of the winds and clouds that run their appointed courses between the sky and earth: there are signs in all these for those who use their minds. (Qur'an 2:164)

Science has set, defined limits beyond which it does well not to comment, lest the utility of the discipline be diluted by false claims and grandiose aspirations. The usefulness of science and its validity should not be questioned, but we also must be well aware of what types of assertions science can make. We must take with a grain of salt every word of the scientist beyond these credible statements. Anything else will have little scientific authority and will come from another mode of thought: probably philosophy.

The above quote from the Qur'an sheds light on the fact that the philosopher and theologian will likely draw much of their insight into the universe from the same source as the scientist. Without divine intervention, we have little recourse to believe things beyond and especially things contradictory to what we see with our eyes in our wonderfully varied and complex world. The philosopher knows as much and carefully perceives phenomena in his world in an attempt to draw out facts and data, much like the scientist.

What many philosophers believe, of which the scientist has no need, is the idea that there are fundamental ideas that pervade the universe and align with logical and physical possibilities. These ideas are not quite the same as scientific law, but rather are less well-defined principles that are inherent to our world and existence.

Such all-pervading ideas can be visualized thusly: "They could see a shaft of light stretching from above straight through earth and heaven, like a pillar, closely resembling a rainbow, only brighter and clearer" (Plato 363). This is a quote from Plato's *Republic*, in which he describes the *Myth of Er*, a religious tale of the afterlife and reincarnation. The light and reality that pierces the planes of existence and unites heaven and earth is centered at what is called the *Throne of Necessity*. This throne unites the different realms in a flourish of reason that defines reality. *Necessity* is one such idea of the philosopher who employs reason to understand the basic tenets of our world.

René Descartes, one of the fathers of modern philosophy, described the source of these unifying concepts as follows:

> It is that, if the intentional reality of any one of my ideas is so great that I am certain that I do not contain this reality in myself either formally or eminently and, therefore, that I myself cannot be its cause, it follows necessarily that I am not alone in the world and that something else also exists, which is the cause of that idea. (Descartes 36)

There are countless ways to understand what Descartes is describing here, and different philosophers have different words explaining what exactly it is. It is clear that scientists and theologians develop their own words to describe the same concept. Many secularists feel that 'Science' is this all-unifying concept. The religious call it 'God'. Philosophers call it many things. But regardless of what it is called, philosophers, theologians, and sometimes scientists too, often describe it using the same term: light.

> I entered and with my soul's eye, such as it was, saw above that same eye of my soul the immutable light higher than my mind... It transcended my mind, not in the way that oil floats on water, nor as heaven is above earth. It was superior because it made me, and I was inferior because I was made by it. The person who knows the truth knows it, and he who knows it knows eternity. Love knows it. Eternal truth and true love and beloved eternity: you are my God. To you I sigh day and night. When I first came to know you, you raised me up to make me see that what I saw is Being, and that I who saw am not yet Being. And you gave a shock to the weakness of my sight by the strong radiance of your rays, and I trembled with love and awe. (Augustine, Confessions 123)

PHILOSOPHY OF LIGHT

The source of all congruency, the source of the mysteriously comprehensive and capable human ability of reason, the source of all good in our world is often conceived of as light. Radiance, brilliance, enlightenment, insight, clarity, illumination, glow. These are all ways we understand the good of our own natural inclination to know and deduce through reason. Such descriptions are riddled throughout philosophical and religious texts across the world, from east to west. Light is inescapable if one seeks to understand either human intellect or the governance of the universe.

We all seem to be drawn to the 'divine light of reason' like moths to flame, and more often than not, human reason is perceived as divine, in the many ways we can understand the word.

> Think how [the soul's love of wisdom and its] kinship with the divine and immortal and eternal makes it long to associate with them and apprehend them; think what it might become if it followed this impulse whole-heartedly and was lifted by it out of the sea in which it is now submerged... Then one really could see its true nature, composite or single or whatever it may be. (Plato 358)

> The irrational separates us, the rational unites us. Thus the immortality of mind or reason is not a personal immortality of separate men, but a share in God's immortality... in so far as men are rational, they partake of the divine, which is immortal. (Russell 172)

These are the opinions of two of history's greatest philosophers, Plato and Aristotle, respectively. As expressed in the above quotes, they certainly both believed, in their time, that the ability of man to consider and determine the rational by means of reason gave man access to, perhaps allowed him to partake in, some divine character. The race of man finds itself united as one rational organism when it allows reason to be its guide.

Neither of these two philosophers would have proposed that such a state has been achieved on earth. Perfect understanding is necessary for the above state to be reached, and perfect understanding requires much more than common man can achieve, "...for a matter is understood when it is perceived simply by the mind without words or symbols" (Spinoza 37). This is reminiscent of our discussion on the nature of *knowing*, and we concluded then that perfect understanding is likely impossible. There are few, if any, situations in which the earth-bound man can say that he has perceived a thing simply as it is, without bias, prejudice, or any perspective at all. Perspective cannot be an aspect of perfect understanding, for naturally, perspective is necessarily subjective and limited; at least the perspective of the eyes, of the senses.

Philosophers have conjectured, however, that we might acquire a perspective removed from the subjectivity of our earthly bodies. Suhrawardī proposed that, "in every seeking soul there is a portion, be it small or great, of the light of God" (Suhrawardi 1). In the act of seeking, we would find a portion of the omniscient, omnipresent perspective of the eternal in ourselves. That perspective would not be held back by any of our dusky mortal constraints.

If the above is possible, the question then becomes how we might expand our access to the divine perspective in order to rid ourselves of our own limited view. Some suppose that we must rid ourselves of earthly concerns. Others feel that we need to develop a pure reverence for the higher things. Still more believe that right earthly action would direct us toward perfect heavenly knowledge. Augustine felt that the human soul must bask in the divine to be illuminated in this way; "the soul needs to be enlightened by light from outside itself, so that it can participate in truth, because it is not itself the nature of truth" (Augustine, Confessions 68). Certainly, not every philosopher shares this opinion of how the soul becomes radiant with understanding, but almost all do agree that there must be an independent source of reason by which the soul comes to know the truth.

Likewise, every philosopher differs in just how we understand the nature of the wisdom that we come to possess. The more scientifically minded will conclude that physical observation and mathematical reasoning are what guide and determine our steps to knowledge. Some rely more on divine revelation. But all philosophers must fall between the two extremes, lest they stop practicing philosophy and start practicing the disciplines of pure theology or science.

> When I say here that I was taught this by nature I only mean that I am led to believe it by some spontaneous impulse and not that I have been shown that it is true by some natural light. There is a big difference between the two. For whatever is shown to me by the natural light of reason... cannot in any way be doubtful, because there cannot be another faculty which I trust as much as that light

and which could teach me that the conclusion is not true. (Descartes
33)

What Descartes is saying here is that the philosophical sense is a higher
recourse for the philosopher than the physical sense. The physical senses are
only accessible to us through the perspective of our physical body, which is
constantly in doubt. He claims that we can be mistaken by natural visions,
but philosophical revelation is less doubtful. However, when the philosopher
finds disparity between the two, he must seek to reconcile the differences, no
matter where the reconciliation leads.

It will not always be the case that human reasoning will overturn
physical experience. Often the data of science will prevail. Einstein
encountered as much when he finally and reluctantly had to accept that the
implications of quantum physics were sufficiently confirmed by science. He
then had to revisit his philosophical sense to determine where it went wrong.
Einstein may have never been able to revise his philosophy, but he was
certainly able to accept the facts that he encountered in the early 1900s. In
as much, he deferred to the portion of his philosophy that placed great
importance on experimental confirmation. He did well to accept what his
bodily senses and mathematical reasoning told him.

If the philosopher achieves his goal of perceiving a thing in its essence,
as it is purely and truly and eternally, as it is defined by its divine character,
he would have seen a magnificent thing indeed. Not all philosophers have
felt that such is attainable. Many modern philosophers would fall into a
category, like most of academia today, rejecting that we can know anything
as it truly is, rejecting the idea that there is such *truth*.

Historically, however, most philosophers did not only propose that
definitive revelation could happen, but most held that they themselves had
seen this pure light and were now trying to communicate in their
philosophical dialogues and treatises what they perceived.

You won't be able to follow me further, not because of any
unwillingness on my part, but because what you'd see would no
longer be an image of what we are talking about but the truth itself,
that is, as I see it; one ought not at this point to claim certainty,
though one can claim that there is something of the kind to see.
(Plato 265)

In this quote, Plato uses careful language, saying that pure and true revelation is possible but qualifying that though he believes he had seen the truth, he cannot claim it as universal certainty. Plato believed that truth was substantial, but something difficult to claim and communicate as one's own.

Augustine likewise lends credence to the idea that the philosopher, scientist, or priest may not have seen the divine truth in and of itself, but he does not limit God's ability to communicate the truth in spite of the individual's faults.

> Even if human insight perceived less than the truth, surely whatever you were intending to reveal to later readers by those words could not be hidden from your good Spirit who will lead me into the right land. (Augustine, Confessions 271)

This communication of divine truth is Augustine's personal idea of the scriptures, but whatever the 'divine source of truth' is, it would communicate unadulterated truth throughout the generations. Scientific, philosophic, or theological truths could all be passed on in this way, directly from the divine source and unaffected by the imperfections of human communication.

The philosopher has a unique concept of this 'divine light of reason' that pertains directly to his own studies, but the concept can be generalized to include the basic facts of the physical world. Light, be it divine or mundane, empirical or philosophical, bears certain qualities toward which almost all life is attracted. Life is drawn to light:

> Because of the correspondence of souls with light, souls flinch from the darkness and are happy in beholding lights. Animals, all of them, seek light in darkness and love the light. (Suhrawardi 135-6)

Suhrawardī could only make this unqualified statement with the unmodern perspective of animal life from which he was writing. Biologists are well aware of much animal life that is completely content in the dark and seek its shelter; however, the twelfth century philosopher's point is still valid for us humans and much animal life. Visible light is still perceived as good by most life, and some light, all along its spectrum, is necessary for all life.

Philosophers and theologians alike take the correspondence of life and light to its extreme in their conceptions of the afterlife, nirvana, and heaven.

PHILOSOPHY OF LIGHT

In descriptions of the glorified human state in the afterlife, they use phrasing such as:

> Every one there is filled full with what we should call goodness as a mirror is filled with light. But they do not call it goodness. They do not call it anything. They are not thinking of it. They are too busy looking at the source from which it comes. But this is near the stage where the road passes over the rim of our world. (Lewis, Mere Christianity 82)

The philosopher Plotinus, who lived in the third century of our common era, described the same phenomenon in his *Tractate on Intellectual Beauty* as Lewis did above, this time in describing the glorified gods of ancient Greece:

> "To these divine beings verity is mother and nurse, existence and sustenance; all that is not of process but of authentic being they see, and themselves in all; for all is transparent, nothing dark, nothing resistant; every being is lucid to every other, in breadth and depth; light runs through light. And each of them contains all within itself, and at the same time sees all in every other, so that everywhere there is all, and all is all and each all, and infinite the glory... all are mirrored in every other. (Russell 295-6)

Plotinus' light is a thing in its essence: beauty or truth or reality in itself. Many philosophers and theologians have agreed with Plotinus' point. For example, there is *beauty* in itself, and only through that beauty are there beautiful things. Thusly, light unites the particular into the universal. Plato, in his roundabout discussion of the perfect government in *The Republic*, commented on pure and simple beauty, in dialogue:

> "And those who can reach beauty itself and see it as it is in itself are likely to be very few."
> "Very few indeed."
> "Then what about the man who recognizes the existence of beautiful things, but does not believe in beauty itself, and is incapable of following anyone who wants to lead him to a knowledge of it? Is he awake, or merely dreaming? Look; isn't dreaming simply the confusion between resemblance and the

reality which it resembles, whether the dreamer be asleep or
awake?"
"I should certainly say that a man in that state of mind was
dreaming."
"Then what about the man who, contrariwise, believes in beauty
itself and can see both it and the particular things which share in it,
and does not confuse particular things and that in which they share?
Do you think he is awake or dreaming?"
"He is very much awake." (Plato 198-9)

To know a thing in itself is a difficult idea for the modern reader to get
her hands around, but it is not impossibly illusive. Still today, when we
consider the development of mathematics and logic, we have to lend some
credence to the philosophers like Descartes, who added much to intellectual
progress in their time.

Math and logic as disciplines are not vital in the current discussion, so
not much attention will be spent on them, save a couple thoughts. These
areas of study and their development are great mysteries for philosophy.
Where exactly the *a priori* ('from what is before') inductions come from, no
one can really say. Many basic mathematical and logical principles seem to
come out of nowhere into the mind of the academic who 'discovers' them. *A
priori* reasoning remains one of the profoundest mysteries in human
knowledge. It is detached from perception and experience, and it is primarily
expressed with symbols and relationships. It is not easy to see these truths
when they are not naturally perceptible; they are difficult to discover.

One such example in logic comes from Spinoza, who went to great
trouble in his work, *On the Improvement of the Understanding*, to produce
definitions and terminology to create a systematic mode based on the *a priori*
arguments of logic. His reasoning, which is difficult to understand but far
more difficult to dispute, establishes logical corollaries based on simple truths:

> I call a thing *impossible* when its existence would imply a
> contradiction; *necessary*, when its non-existence would imply a
> contradiction; *possible*, when neither its existence nor its non-
> existence imply a contradiction, but when the necessity or
> impossibility of its nature depends on causes unknown to us, while
> we feign that it exists. If the necessity or impossibility of its
> existence depending on external causes were known to us, we could
> not form any fictitious hypothesis about it; whence it follows that if

there be a God, or omniscient Being, such a one cannot form fictitious hypotheses. (Spinoza 233)

In his last clause in the above quote, Spinoza once again lends credence to what many of his contemporaries and predecessors said: namely, if there is something that could be conceived of as 'the divine', its knowledge of any given thing and all things at once would be perfect and based solely on those things as they truly and completely are, not on relationship and perception. Divine knowledge would be based on *a priori* reasoning, pure from the source.

The question for many then becomes, "Is there such a source?" Spinoza and other secularists may suppose that nature and nature's laws are that objective reality, but probably inaccessible to humanity in their totality. Others with more spiritual optimism would say that such a view exists, but access to it is guarded by divine law, around or through which very few have peered. Either way, the *a priori* reasoning of logic and mathematics is supremely illusive and mysterious for us mortals, and only the truly blessed ever discover new truths of this kind.

If these slippery considerations and ideas are distasteful for the empiricist, let him or her remember that serious philosophers are compelled by similar forces by which the scientist is restricted. Philosophers also must observe before they believe, though that observation will look very different from the scientific. Only after perceiving with the senses, be they physical or philosophical, can a thinker internalize facts and truths.

> Let us class together as "faculties" the powers in us, and in other
> things that enable us to perform all the various functions of which
> we are capable [i.e. vision, hearing, reasoning]... I can only identify
> a faculty by watching its field and its effects [i.e. sound, light, truth].
> (Plato 200)

We only know the faculties of our senses by observing the effect that the world has on those faculties. We would know nothing of our sense of vision if we did not have light to interact with it. So too, we know nothing of the validity of human reasoning without truths to test it. If for no other purpose, we must hold onto the possibility of finding truth so that we do not rid ourselves of the necessity of our reason. Reason is a powerful thing and requires truths to seek to remain sharp and active and real.

THE GLORY OF MAN

> But what, then, am I? A thinking thing. And what is that? A thing
> which doubts, understands, affirms, denies, wills, does not will, and
> which also imagines and senses. (Descartes 26)

The above is the basis of Descartes' most famous philosophical concept,
Cogito Ergo Sum: 'I think therefore I am'. In his time and still in ours, there
are those who question what we can know and go on to propose that we can
know very little, if anything. We have access to the intellectual gymnastics
that would allow us to question our very existence. Movies, such as *The
Matrix* and *Inception*, deal with this idea. What is real? Is anything real? Is
this all a dream? Descartes rejected such notions as silly. He developed parts
of his philosophy to contradict the thought that nothing is knowable.

Descartes says that we know that we exist because we can consider the
fact whether or not we exist. An entity without existence could not do as
much. And if we do exist and have the ability to consider the fact, a cascade
of logical consequences falls into place.

One part of the essence of human existence, at least in Descartes' view,
is that humans possess *will*. Free will is a hot topic in Christian theology and
a point of contention in philosophy besides, but Descartes and other Catholic
philosophers took it for granted. Their human experience told them that they
constantly had the very present opportunity to will one thing or will another,
to desire and do one thing or another.

Will is at the heart of many philosophies and theologies, and it will play
a major role in our current discussion. Whether or not it exists, whether or
not it has effect and consequence in our world, we will explore in depth. It
is the heart of the divine light; as Descartes and others would suggest, it is the
heart of the man.

> As in water face reflects face,
> > so the heart of a man reflects the man. (Prov. 27:19)

Solomon, the writer of the biblical book of Proverbs, seems to believe
that the inner life of an individual and all the complications of his or her
aspirations, motivations, and choices, are what show the outer world who the
person actually is. The essence of the man is the heart of the man, and it is
difficult to understand the heart of man without agency to express it. Plato

too suggests that there are three primary elements in each human: reason, appetite, and spirit. Again, the combination of these things seems to be pointing the reader to something akin to an active, expressive will.

To see the positive quality of the will of man, perhaps it is best to see what the thing perverted looks like. Augustine considers the fallen angels and condemned human souls in a statement to God:

> By the wretched restlessness of fallen spirits, manifesting their darkness as they are stripped naked of the garment of your light, you show how great a thing is the rational creature you have made. (Augustine, Confessions 277)

In the corruption of the good will, we see the powerful influence of man's glory in the world. A quick survey of a newspaper or newsfeed on Twitter would inform anyone what kind of impact an evil will can have on all of us. Terrorism and senseless acts of violence are perpetrated by individuals who have an aspect of the divine inside them, twisted into some horrible thing. We can hardly even recognize where such hatred comes from, but the power of that hate bears the same qualities of the power of the best good of our earthly experiences. That power again and again brings tears into our eyes, be they of joy or sorrow.

But will and human reason can be imperfect without being completely perverted from the divine good. Descartes goes on to consider our worldly condition and the will:

> If, for example, I consider my faculty of understanding, I recognize immediately that it is very limited and finite and, at the same time, I form the idea of another similar faculty which is much greater – in fact, the greatest possible, and infinite – and from the mere fact that I can form this idea I perceive that it belongs to the nature of God... I experience the will alone, or freedom of choice, as being so extensive in my own case that I conceive the idea of none greater, so that it is principally because of this faculty that I understand myself as being in some sense the image and likeness of God. (Descartes 47)

Descartes is so wed to the concept of human will that he cannot imagine anything greater in us, and insofar as will is the greatest faculty in man and man has been created (in the Judeo-Christian understanding) in the image of

God, Descartes relates the human will back to its divine source. Many folks will find this proposal difficult and maybe impossible to accept. I would ask, however, for this reader to be patient with my treatment of the will. Only in the balance of the text will the idea be fully conceptualized.

In one of the more respected texts of Buddhism (though there is no single 'holy' scripture in this eastern faith), the idea of the will is defended simply by stating, "You are the master of yourself. What other master could there be?" (Dhammapada 12:160). Surely there could only be one other, the divine, and each has to consider whether such a one allows independent beings of will or if the divine will is the only will in the universe. Augustine answers the question as follows:

> For it is he who made the world, filled with all good things, things accessible to sense, and those perceived by the understanding; and in the world his greatest work was the creation of spirits, to whom he gave intelligence, making them capable of contemplating him, able to apprehend him... He has bestowed on these intellectual natures the power of free choice, which enabled them, if they so chose, to desert God... and yet he did not deprive them of this power, judging it an act of greater power and greater goodness to bring good even out of evil than to exclude the existence of evil. There would not, in fact, have been any evil at all, had not that nature which was capable of change (although good and created by the supreme God who is also the changeless good, who made all things good) produced evil for itself by sinning. This sin is itself the evidence that proves that the nature was created good; for if it had not itself been a great good, although not equal to the Creator, then assuredly this apostasy from God, as from their light, could not have been their evil. (Augustine, City of God 1022-3)

Augustine asks two questions in the above quote, one directly and one indirectly. First, he wonders what evil could be unless it is a turning away, a perversion, of the divine will. Naturally, the only way for a perversion of that will to occur is for there to be such a possibility, and it is impossible for the changeless and perfect divinity to thus pervert its own will. This implies Augustine's indirect assertion, that there must necessarily be other wills which affect the sinful turn. Inasmuch as another will affects this perversion, will and will alone is the creative force of sin in an otherwise perfectly good universe created by a perfectly good divinity.

> I was brought up into your light by the fact that I knew myself both to have a will and to be alive. Therefore when I willed or did not will something, I was utterly certain that none other than myself was willing or not willing. That there lay the cause of my sin I was now coming to recognize. (Augustine, Confessions 114)

Augustine, in his own practical experience and history of questionable behavior, recognized himself as the culprit producing sin, and he thus laid the blame on no one else. He or she who would not allow for human will must find another source of his or her sin (if such a thing exists as sin), or generally, a source of his or her choices. Without another source to blame for a person's action, the blame will be as inescapable as the human will.

Wittgenstein likewise recognized the ubiquity of will in our lives. He stated, "But in the sense in which I cannot fail to will, I cannot try to will either" (Wittgenstein, Philosophical Investigations 161e). The will is inescapable and active, and one begins to question if it can be controlled at all. If we cannot 'try' to will, is the will and everything besides controlled by a sovereign divinity? The question must be stopped dead by making the distinction between our inability to *not will* and our responsibility to control our will. The will is relentlessly present, always hand in hand with consciousness, always accompanying our consciences. A sovereign foot is on the gas of the present, and the will is the steering wheel. It is always turning one way or another, even when seemingly inactive. We *will* because the present incessantly compels us to.

The patient and dedicated reader who sought another perspective on the complexity of humanity's relationship with the will would do well to read John Steinbeck's magnus opus *East of Eden*. The entire narrative of the lengthy novel focuses on individuals in one family and how they all make their own choices and are responsible for those choices, regardless of their situations and influences and genetics.

The portion of Steinbeck's novel with which we are concerned at present is found in Chapter 24. Three characters, Samuel, Adam, and Lee, are discussing the biblical story of Cain and Abel. Samuel is a gritty character who represents the earthly wisdom of the American West, and Lee is Adam's Chinese servant who represents the mystic wisdom of the Far East. Lee had engaged Confucian scholars to study the story of Cain and Abel in an attempt to understand it fully.

The three men discussed the story before and went away dissatisfied with their interpretation, so when they again reflected on the topic, Lee was excited to offer new insight. The verses that so troubled them are Genesis 4:6-7, where YHWH (the deity of Judaism and the Christian 'God' of the Old Testament) charges Cain, in view of the killing his brother Abel, to resist future sin. The current English Standard Version translates it as follows:

> If you do well, will you not be accepted? And if you do not do well, sin is crouching at the door. Its desire is for you, but you **must** rule over it. [bolded for emphasis]

It is with these verses that the following dialogue is concerned. Lee begins by telling the long process his scholars went through to develop their own translation of the Hebrew text. What follows is a lengthy portion of the novel:

> "After two years we felt that we could approach your sixteen verses of the fourth chapter of Genesis. My old gentlemen felt that these words were very important too – 'Thou shalt' and 'Do thou.' And this was the gold from our mining: '*Thou mayest.*' 'Thou mayest rule over sin.' The old gentlemen smiled and nodded and felt the years were well spent. It brought them out of their Chinese shells too, and right now they are studying Greek."
>
> Samuel said, "It's a fantastic story. And I've tried to follow and maybe I've missed somewhere. Why is this word so important?"
>
> Lee's hand shook as he filled the delicate cups. He drank his down in one gulp. "Don't you see?" he cried. "The American Standard translation *orders* men to triumph over sin, and you can call sin ignorance. The King James translation makes a promise in 'Thou shalt,' meaning that men will surely triumph over sin. But the Hebrew word, the word *timshel* – 'Thou mayest' – that gives a choice. It might be the most important word in the world. That says the way is open. That throws it right back on a man. For if 'Thou mayest' – it is also true that 'Thou mayest not.' Don't you see?"
>
> "Yes, I see. I do see. But you do not believe this is divine law. Why do you feel its importance?"

"Ah!" said Lee. "I've wanted to tell you this for a long time. I even anticipated your questions and I am well prepared. Any writing which has influenced the thinking and the lives of innumerable people is important. Now, there are many millions in their sects and churches who feel the order, 'Do thou,' and throw their weight into obedience. And there are millions more who feel predestination in 'Thou shalt.' Nothing they may do can interfere with what will be. But 'Thou mayest'! Why, that makes a man great, that gives him stature with the gods, for in his weakness and his filth and his murder of his brother he has still the great choice. He can choose his course and fight it through and win." Lee's voice was a chant of triumph.

Adam said, "Do you believe that, Lee?"

"Yes, I do. Yes, I do. It is easy out of laziness, out of weakness, to throw oneself into the lap of deity, saying, 'I couldn't help it; the way was set.' But think of the glory of the choice! That makes a man a man. A cat has no choice, a bee must make honey. There's no godliness there. And do you know, those old gentlemen who were sliding gently down to death are too interested to die now?"

Adam said, "Do you mean these Chinese men believe the Old Testament?"

Lee said, "These old men believe a true story, and they know a true story when they hear it. They are critics of truth. They know that these sixteen verses are a history of humankind in any age or culture or race. They do not believe a man writes fifteen and three-quarter verses of truth and tells a lie with one verb. Confucius tells men how they should live to have good and successful lives. But this – this is a ladder to climb to the stars." Lee's eyes shone. "You can never lose that. It cuts the feet from under weakness and cowardliness and laziness."

Adam said, "I don't see how you could cook and raise the boys and take care of me and still do all this."

"Neither do I," said Lee. "But I take my two pipes in the afternoon, no more and no less, like the elders. And I feel that I am a man. And I feel that a man is a very important thing – maybe more important than a star. This is not theology. I have no bent toward gods. But I have a new love for that glittering instrument, the human soul. It is a lovely and unique thing in the universe. It

is always attacked and never destroyed – because 'Thou mayest.'"
(Steinbeck 301-2)

Lee, in understanding the verb with which YHWH commanded Cain, found an unshakable belief in and appreciation for the human soul, the human will. The ability to choose. The responsibility to choose. *Timshel. Thou mayest.* For him, and for the current reader, this is not necessarily a theological concept; it is purely philosophical. Man has the ability and responsibility to choose, with or without any other divine thing in the universe, and inasmuch, man's will *is* a sort of divine thing.

The question that this text will move toward presently is whether another greater divinity exists in our reality to which man's will should submit. Augustine speaks of this divinity, and man's divine qualities, and man's slavery to sin or righteousness in his own interpretation of the same verses as above in his own magnus opus, *City of God.*

> Thus, when God says, 'For (there is to be) a return of it to you,' the verb to be understood is 'should be' rather than 'will be'; it is said by way of prescription rather than prediction. For a man will have mastery over his sin if he does not put it in command of himself by defending it, but subjects it to himself by repenting of it. (Augustine, City of God 605)

Philosophically, theologically, and, as we will find indirectly, even scientifically, the will is an extremely important aspect of the universe, but the question remains: 'Is there something that might direct this will?' Philosophy has an answer to the question, which we will explore next in the depths of Plato's cave.

PLATO'S CAVE

Many ancient religions and philosophies can be classified as monism. Monism is any system of thought or devotion conceiving the universe as primarily, perhaps entirely, controlled or created by one essence, be it a god or a good. Conversely, in a dualist universe, there are two powers that compete in creation and control, often striking some sort of cosmic balance. Dualism can be used to understand the ancient religion of Zoroastrianism and the Chinese philosophy of yin and yang. Whether the two sides of these dualisms have one ultimate source is immaterial. Dualists would maintain that the two divine characteristics are completely separate and competing in our physical world.

Monist philosophies, which have dominated the west for millennia, include the philosophies of many of history's most respected thinkers, and monist religions are among the most practiced in the world, including Christianity, Judaism, and Islam. Christians and Muslims together compose half of the world's population, so most people must identify as monists. For these religions there is one absolute divinity: God, YHWH, or Allah; and to this one god all creation owes its existence.

The major intellectual difficulty of monism is typically centered around the presence of evil in the world and its source, whereas good and nature and law are easily attributable to the one god directly. Dualism does not have trouble with this issue, as good and evil are seen as two equal and independent forces at play in our universe, but monism must conceive of a source of evil unlike and separate from the ultimate and divine goodness.

Since all monists agree on the existence of one ultimate, divine source, it is not necessarily incorrect to say that they are all worshipping the same god. This is not as outlandish a statement as it may appear at first blush. The religiously devout of one or another sect may not like the fact, but the disgruntled must say that though they all worship one god, the others worship sacrilegiously.

> The only disagreements among the monists concern the attributes of the One, not the One itself. Since the One is the source of all things and includes all things in it, it cannot be defined in terms of those things, since no matter what thing you use to define it, the thing will always describe something less than the One itself. The One can only be described allegorically, through the use of analogy, of figures of imagination and speech. (Pirsig 498)

We all call this monist divinity what we like, but the names often bear little distinction apart from the language used. "It will often prove useful in philosophy to say to ourselves: naming something is like attaching a label to a thing" (Wittgenstein, Philosophical Investigations 7e). A name is usually nothing more than something we can refer to later to ensure that we are still discussing the same thing. The description and connotations of the name usually do not bear much meaning in themselves. It is only in the *divine whole* that we have our conception of the one god, and no one can honestly claim to have access to that whole by which a full description could be made. Even more, our imperfect human language could not contain the whole even if we did perceive it.

So instead, what happens is that everyone simply uses different words that do the best to describe their idea of the *one*. Philosophers do the same.

In Robert M. Pirsig's hugely successful modern philosophical work, *Zen and the Art of Motorcycle Maintenance*, the narrator gives a full treatment of this concept in describing his circuitous path to philosophical enlightenment. Pirsig's narrator gives a nod to philosophy's past terms that were used to describe the *one*. Philosophers have called it good, beauty, righteousness, rightness, nature, etc., etc. Plato preferred the Greek term *aretê*, which translates as *virtue*, while Pirsig landed on the term *quality*. Regardless of the term used, the philosopher who employs it will admit that the word hardly encompasses the full meaning he intends. We humans understand many of these words intuitively and would have difficulty in giving a fair definition of the *one*, but regardless of our intellectual or philological failings, we attribute all goodness of the pure, full meaning to the *one*.

In truth, philosophers cannot even get their minds around what exactly oneness means in our world. Aristotle mused, "Are goods one, then, by being derived from one good or by all contributing to one good, or are they rather one by analogy?" (Aristotle 9). In one way or another, all three of Aristotle's options probably contribute to the reality of the form of the *one*. We do not understand how a good world created by the one ultimate good contains various good things of different qualities.

Philosophers and theologians use the *one* as not only the source of good, but also the end of all. That which humans should seek without qualification or consideration is this *one*. It is the means and the ends.

In light of this fact, Aristotle took a pragmatic, yet clearly selfish approach in saying, "Happiness, then, is something final and self-sufficient, and is the end of action" (Aristotle 11). For Aristotle, the ultimate end of our present action was not good or virtue or quality, but one's personal happiness. Of course, many factors should be considered in what ultimate happiness is, but this was Aristotle's *one*.

A paradigm shift of some magnitude would be necessary for generations to start to conceptualize their one good or god by some other means, though individuals may experience rapid and sometimes violent changes in themselves by which they come to understand the *one* differently. Religious conversion is an easy example of this process: the one god comes flashing into view where it was not seen before. Conversely, apostates from any given religion may seek shelter in the absolute ends of science, or, perhaps more likely, will reject monism altogether and lean toward agnosticism.

With a sufficient understanding of monism, we are now prepared to discuss one of the most well-known philosophical thought experiments and allegories, *Plato's Cave*, but before we delve into his cave, let us take a look at another cave found in the Qur'an.

In Sūrat al-Kahf, the eighteenth chapter of the Qur'an, Muhammad describes a cave in which devout youths hide from the wrath of disbelieving townsfolk. An unknown number of youths and their dog went into the cave which was completely black with darkness, and Allah gave them a supernatural sleep. The youths stayed sleeping in the cave for an indeterminate period of time, and when they woke, they themselves did not know how long they had been sleeping. When one went into town to buy provisions, the townsfolk did not know how long the youths slept in the cave or even how many there were. The truth was impossible to apprehend, except that Allah would illuminate the truth, for any physical evidence was impossible to collect in the supernaturally dark cave.

Muhammad uses this cave as an example of Allah's sovereignty of knowledge, followed by a story in which Moses lacks the prophetic foresight that an angelic guide has through Allah. Allah knows that he keeps the world in darkness, and only at distinct times through specific prophets does he illuminate profound truths that accompany the mundane truths found expressed in nature. Many times in the Qur'an, the mundane is argued to be proof of Allah's existence and goodness, but any doubter could attribute the same proofs to any source he likes, as Muhammad makes abundantly clear concerning disbelievers. It is not easy to follow the argument in this sura, as it purposefully lacks key information known only to Allah, who like a divine Sherlock Holmes is happy to expose all at the end, at the day of judgement.

Plato's Cave, on the contrary, gives a picture of how truth is illuminated in the darkness of our mortal state, but only in part, which gives us hope to fully know truth through general revelation and mature reason.

In Plato's allegory, the light from the outside world is directly tied to vision and human reason. We find as light is to vision, so good is to reason.

Plato felt, as most philosophers do, that the knowledge of a thing in itself is more perfect than the idea of the thing. The mortal concepts of the truths we witness indirectly in our physical existence are nothing more than shadows cast into a world of ignorance and inexperience. A direct view of light is better than the shadows it creates.

> I want you to go on to picture the enlightenment or ignorance of our human condition somewhat as follows. Imagine an underground chamber like a cave... In this chamber are men who have been prisoners there since they were children, their legs and necks being so fastened that they can only look straight ahead of them and cannot turn their heads. Some way off, behind and higher up, a fire is burning, and between the fire and the prisoners and above them runs a road, in front of which a curtain-wall has been built, like the screen at puppet shows between the operators and their audience, above which they show their puppets. (Plato 241)

All their life, these men would experience no vision except what is shown on the wall in front of them, cast by shadows of that behind. There is no reality for them by movement or perception of a thing in itself, but only the indirect actions of the shadows on the wall.

> And so in every way they would believe that the shadows of the objects we mentioned were the whole truth. (Plato 241)

These two men would know nothing else. The shadow world on the wall would be their reality. Even concepts of the self would be blurred and distorted in the shadow.

At some point in their adult life, after these men have been in such a situation since childhood, one is somehow freed of his bonds and able to explore the cave and beyond into the outside world. When he turns away from the wall and sees something not quite like all that he had ever experienced, he would be astounded. He would exit the cave to see these new and wonderful things.

> When he emerged into the light his eyes would be so dazzled by the glare of it that he wouldn't be able to see a single one of the things he was now told were real. (Plato 242)

The man who had only ever known dusky shadows would hardly be able to perceive objects illuminated by natural day light, and he would be tempted to retreat back into his comfortable reality in the cave. In fact, he would have great difficulty in believing that these things that he sees in the light are real. His reality has always been of a different sort, and the world outside the cave would be less real for a time.

But having seen and experienced things as they truly are, the man comes to a point where he does not at all wish to return to his prior state. The reality that he now possesses is glorious and tangible in ways it could never be in the cave. In fact, the man would wish to return to the cave and release his companion so the other could experience this more real reality.

The first man, however, would have difficulty explaining what he experienced to the man who still has only ever known shadow. He must try to explain the dimensionality and complexity of the real world using only the two-dimensional terms the second man can understand. Surely, the description would lack the quality that the first man wished to communicate. How could he share more than the simplicity of the shadows in the cave world when he had no common language to bridge the gap? Furthermore, by stepping in front of the second man in his real three-dimensional form, he might terrify his companion with his own dimensionality.

In Plato's allegory, the sun is the ultimate divine good, by which everything receives its good quality. In seeing the things of the earth as they are in the light, the freed man would witness that quality, but he would not see the one good in itself. Only after much time adjusting his eyes to the brightness of his new reality would the man be able to turn his sight to the source of universal goodness, the sun.

The final thing to be perceived in the intelligible region, and perceived only with difficulty, is the form of the good; once seen, it is inferred to be responsible for whatever is right and valuable in anything, producing in the visible region light and the source of light, and being in the intelligible region itself controlling source of truth and intelligence. (Plato 244)

Apply the analogy to the mind. When the mind's eye is fixed on objects illuminated by truth and reality, it understands and knows them, and its possession of intelligence is evident; but when it is fixed on the twilight world of change and decay, it can only form opinions, its vision is confused and its opinions shifting, and it seems to lack intelligence. (Plato 234)

So, Plato has a monist belief in a supreme good, the form of good itself, which gives divine character to our earthly world. The good, 'virtue' in Plato's *The Republic,* is what illuminates the realities of the world and allows the man of right action and opinion to skillfully navigate a difficult and shadowy existence in this life. Plato's good is truth. It is right. It is philosophical revelation. This allegorical conception of light is shared by most of the world's religions.

Truth is the Sun's extended light. (Rig 1:105.12).

I enjoyed reading [philosophical books], though I did not know the source of what was true and certain in them. I had my back to the light and my face towards the things which are illuminated. (Augustine, Confessions 70)

The wisdom which itself needs no light, illuminating needy minds, the wisdom which governs the world down to the leaves that tremble on the trees. (Augustine, Confessions 117)

I believe in Christianity as I believe that the Sun has risen, not only because I see it, but because by it I see everything else. (Lewis, Is Theology Poetry 140)

This is monism. This is much of western philosophy and theology. There is one source of good and existence in the universe, by which humankind is able to experience all else. It might be philosophical good or virtue or beauty or quality. It might be theological divinity: God or Allah or YHWH. It might be scientific law or the quantum character that defines all matter and energy. Regardless of who is talking about it and from what perspective, we find time and time again the ends of monism defined in terms of light. It seems that mankind cannot understand reality at all but that we understand it as light.

Light is our reality. Essentially all religions across the entire world acknowledge the same in their sacred texts and rituals. In the next chapter, we will begin to consider what the world religions have to say about the theology of this ultimate good, this divine light.

THEOLOGY OF LIGHT

Theology attempts to explain the realities of our human experience much in the same way philosophy and physical science do. It is no stretch to say that the practices are very different, but as in the other disciplines, theologians believe in certain abiding universal laws that direct our understanding of the world in systematic ways. The religious explanation will likely employ mysticism, as do certain philosophical and scientific understandings, but unless everything is explained with reasonable comprehensiveness and uniformity, the religiously devout have no responsibility to accept what the priest is teaching. Religion should not and cannot abolish reason.

Unless religion teaches what is intellectually acceptable, it will only ever attract the fundamentally extreme perspectives of those completely detached from reality. Many of the most intelligent people in all of human history believed in the divine, and they balanced their belief with reason. The modern thinker does well to not dismiss well-entrenched, historical belief with a flourish of scientific materialism. Theology requires as careful consideration as philosophy and science.

Sagan fell into the trap of belittling religious thought, as previously discussed in the second chapter. Again, from *Pale Blue Dot*, he asks, "But if the Bible is not everywhere literally true, which parts are divinely inspired and which are merely fallible and human?" (Sagan 42). Here the renowned physicist asks a question that has haunted man for centuries. How are we to trust in the divine inspiration of holy scriptures if we cannot interpret every word and phrase literally in every chapter and verse? Literary devices like poetic description, metaphor, and allegory must be interpreted. How can the divine inspire words that will necessarily be left to the reader to interpret imperfectly for herself? Further, how are any of the countless translations of the original texts any good in the face of literalism? How can the gods be so flippant with human devotion?

Augustine had a method for dealing with this interpretive difficulty. He felt that God would not begrudge the bishop the use of his God-given intelligence: "I was ordered to believe Mani. But he was not in agreement with the rational explanations which I had verified by calculation and had observed with my own eyes" (Augustine, Confessions 75). Augustine considered his old sect's teachings on astronomy and astrology and found that the conclusions of that religion and philosophy did not align well with what he perceived in reality. When he saw this disparity, Augustine's decision to leave Manichaeism was obvious. He must betray that which betrayed his reason. Augustine went on to find a comfortable home in the Catholic faith, which made no verifiably erroneous claims on the physical world as did Manichaeism. Augustine left a religion that had grown repulsive to him for a religion that seemed to address the questions about which the philosopher demanded reasonable answers.

When the nomenclature *religion* or *theology* is applied to any thought or idea, we all have an automatic reaction that is typically positive or negative, based on our experience of religion. If one was told to ignore biology and evolution while being slapped on the knuckles by a nun with a ruler, naturally he will have certain visceral reactions when discussing either religion or science. It is nigh impossible to remain unbiased in the face of topics of emotional or psychological difficulty, but we must if we hope to maintain intellectual composure and to address the issues that need addressing.

> What sensitive ears ordinary people have... in matters of religion! They cannot tolerate the discussions of philosophers about the immortal gods. Yet they not merely tolerate, they listen with pleasure to fictions, sung by poets and acted by players, which offend against the dignity and the nature of the gods. (Augustine, City of God 235)

In this quote, Augustine laments man's ability to isolate the idea of the divine. We are too often able to treat with pleasure the absurdities proposed in Greek dramas and comedies as they are now portrayed in vulgar satires like the long-running animated sitcom, South Park. But when it comes to serious discussion, the typical secularist is likely to belittle religion to the point that he completely disdains to talk about it at all. Religion offends, but only when it is earnestly believed. The secularist can hardly muster any respect for earnest religious devotion; whereas, he can mock relentlessly the religiously indifferent.

Naturally, the discussion is a difficult one. On one hand, those opposed to discussing theology seriously impose requirements upon the religious, asking for philosophical and scientific arguments on which to base the validity of religion. On the other hand, when the discussion becomes complicated by theological enigma and paradox, the skeptic will decry religion for being overly convoluted and impossibly complex. But,

> If we ask for something more than simplicity, it is silly then to complain that the something more is not simple. Very often, however, this silly procedure is adopted by people who are not silly, but who, consciously or unconsciously, want to destroy Christianity. Such people put up a version of Christianity suitable for a child of six and make that the object of their attack. When you try to explain the Christian doctrine as it is really held by an instructed adult, they then complain that you are making their heads turn round and that it is all too complicated and that if there really were a God they are sure that He would have made 'religion' simple. (Lewis, Mere Christianity 31)

This is the view that Sagan takes by demanding that the Bible be interpreted literally while at the same time making fun of literal biblical interpretations. The conversation does not even begin before it is quashed by the impossible demands of some devotees of atheism and agnosticism or, more troublingly, of science.

However, when scripture is interpreted with a certain flexibility, logical absurdities become reasonable enigmas that rival the profundity of the wave-particle duality of light. Scripture is no less difficult for these complexities, but it is not killed by logical falsifiability.

> Above all, I heard first one, then another, then many difficult passages in the Old Testament scriptures figuratively interpreted, where I, by taking them literally, had found them to kill. So after several passages in the Old Testament had been expounded spiritually, I now found fault with that despair of mine, caused by my belief that the law and prophets could not be defended at all against the mockery of hostile critics. (Augustine, Confessions 88)

Augustine, a man of supreme intelligence, railed against the silly doctrines of the Old Testament while he remained a follower of Mani, but after being repulsed from his old religion by its intellectual inconsistencies, Augustine found that figurative interpretation of the ancient scriptures imbued vitality where previously were only dead words. Likely every text from every religion must necessarily be interpreted figuratively in at least some instances. Most scripture employs poetry and poetic descriptions of the divine, which one would be seen as ridiculous to interpret literally. No student of literature would recommend this kind of reading on any poetic text. Interpretive flexibility is key, even when one views the text as the inspired word of the divine.

Thus, C.S. Lewis recommends that Christians be unworried by the harrying mockery of unbelievers. What they mock is not at all what serious Christians believe:

> There is no need to be worried by facetious people who try to make the Christian hope of 'Heaven' ridiculous by saying they do not want 'to spend eternity playing harps'. The answer to such people is that if they cannot understand books written for grown-ups, they should not talk about them. All scriptural imagery (harps, crowns, gold, etc.) is, of course, a merely symbolical attempt to express the inexpressible. (Lewis, Mere Christianity 76)

Naturally, religious people should not be worried by such attacks from the unbeliever. However, the religious are often offended by these attacks because the above is exactly what they do believe. An elementary reading of scripture will only allow an elementary understanding. And an elementary understanding of the divine will find itself losing ground in intellectual combat, even with the intellectually simple. Christians assert that individual faith can and should be as simple as a child's, but maintaining a certain theology in an intellectually hostile world will require refinement of that theology and a number of serious religious thinkers to defend it.

Most theologians believe, luckily, that the divine does not trifle with his followers by hiding basic truths from their senses. The divine character and its defense ought to be clearly demonstrated in the world of our daily experience.

> There are things of which the knowledge is fixed and determined without evolving with the generations, such as the lights of wisdom and knowledge. But while the truths of these things remain the same, their embodiments in the physical realm are both many and

varied... [God has] relieved the tedium for mortal senses by the fact
that what is one thing for our understanding can be symbolized and
expressed in many ways by physical movements... These physical
things have been produced to meet the needs of peoples estranged
from [God's] eternal truth." (Augustine, Confessions 288-9)

Here Augustine lends credence to the concept that God and his truths
are unchangeable but that his truths may also be expressed in our
understanding of nature in different ways. God's truths will thus adapt their
expressions of his character to the era in which they are operating for the
benefit of his people and those who do not yet believe. The divine is not
limited to one time or one interpretation, but its truths will be accessible to
all people in all ages and all regions, people of various intellects and
philosophical bents. The plurality of interpretation through time is just one
way that the divine can help us mortals. Likewise, plurality of interpretation
of one text in one time for one person exposes and explains profound enigmas
through those interpretations, and it is a powerful tool for the theologian
attempting to understand the divine character.

> By this blessing I understand you to grant us the capacity and ability
> to articulate in many ways what we hold to be a single concept, and
> to give a plurality of meanings to a single obscure expression in a
> text we have read. (Augustine, Confessions 296)

> There is something to be gained from the obscurity of the inspired
> discourses of Scripture. The differing interpretations produce many
> truths and bring them to the light of knowledge. (Augustine, City
> of God 450)

> See now how stupid it is, among so large a mass of entirely correct
> interpretations which can be elicited from those words, rashly to
> assert that a particular one has the best claim to be [the real one],
> and by destructive disputes to offend against charity itself.
> (Augustine, Confessions 265)

Augustine, perhaps the greatest Christian thinker of all time and among
the preeminent philosophers from all ages, felt that the obscurity and enigma
found in the biblical texts did not at all detract from his understanding and
acceptance of the theology communicated therein. In fact, the philosopher

felt that the obscurity and enigma actually added to and perfected his understanding of reality and the workings of his world. Without any kind of view toward the scientific paradoxes and profound truths that twentieth century science would expose, Augustine seems to have preempted the ways in which the modern secularist would understand the workings of the world.

Perhaps it is the case that scientific and theological enigmas with which philosophers currently have to contend will soon become clear through a deeper understanding or a complete revision of our limited perspective. Perhaps nature and reality do not trifle with such difficult complexities. Perhaps enigma is just a feature of incomplete knowledge.

If that is the case, we all have a far way to go. If not, let us accept enigma wherever it surfaces and not use complexity of reasoning as criteria by which we can delegitimize a certain philosophy, religion, or science.

Religious knowledge is a unique sort of intellectualism in our world. The religious man believes in things that he has not necessarily seen or experienced, not based on the authority of peer-reviewed, systematic studies, but based on the authority of a religious figure who claims to have access to divine truths inaccessible to the layman. This mode of acquiring faith is dealt with in part by scripture's order to test a prophet's claim to truth by establishing his predictive power. The theologian's process of verification is remarkably akin to the ways in which we accept theoretical physics. But the testing of a prophet and his truths must be done seriously, disregarding the abstractions of astrology and fortune telling, by aligning the prophet's claims directly with historical, physical, and personal fact.

If the prophet's claims can be verified by such facts and aligns with the rest of the divine scripture, which has likewise been confirmed, the religious man has a legitimate source of belief in such a prophet and prophecies. He can hold onto his convictions tightly, perhaps more tightly than the scientist to his truths. But when new evidence for or against his point of view arises, he cannot ignore it. When new scientific discoveries come to light, the theologian must contend with them. He must not totally disregard what is yet believed by other disciplines. His narrow concept of the divine cannot alone be proof against otherwise valid hypotheses from science, philosophy, or other religions, for his imperfect understanding of the divine should daily be reinterpreted to fit a new worldview. Longstanding truths and personal experience should not be abandoned, but interpretations can adapt to fit with new revelation from other sources of human knowledge, as they should.

For the Christian specifically, Lewis proposes that truths from outside of the faith should be internalized by the faith in order to constantly be growing the breadth of the faith accessible to the believer of the Christian God.

> If you are a Christian you do not have to believe that all the other religions are simply wrong all through... If you are a Christian, you are free to think that all those religions, even the queerest ones, contain at least some hint of the truth... But, of course, being a Christian does mean thinking that where Christianity differs from other religions, Christianity is right and they are wrong. (Lewis, Mere Christianity 29)

Lewis would recognize the text of the Bible as what is 'right', but again, human interpretation is often faulty. Scientific theories have been disproven many times throughout the centuries, but the authority of yet unknown 'natural laws' have not at all been questioned. So too, scriptural interpretation can be revised without scriptural authority bearing any responsibility for the shortcomings of the specialists trying to understand it.

> All scripture is breathed out by God and profitable for teaching, for reproof, for corrections, and for training in righteousness, that the man of God may be complete, equipped for every good work. (2 Tim. 3:16-17)

From Lewis' sacred text itself, the Bible, we find the valuable perspective delineated above. The apostle Paul wrote the above in a letter to his dear friend and mentee, Timothy. He wrote a personal letter to his personal friend. At the time, certainly, Paul could not possibly have conceived such a letter finding its way into the canon of the holy scripture of his faith. In this particular quote, Paul was referring only to the Old Testament as it is now understood, not the gospels, epistles, and prophecies of the New Testament, though the modern Christian reader includes the New Testament as likewise being 'God-breathed'. But when the New Testament did not yet exist, the apostle could not have considered it, and when he was writing such a letter, he could not have presumed that his own words would be canonized.

Even so, when Paul refers to the Old Testament, he is referring to various scriptures written by mortal individuals. The writers of Chronicles believed themselves to be writing the history of Israel and Judah, not scripture, as did the writer of the Pentateuch. David wrote personal poems for YHWH, perhaps even as corporate worship for the nation he ruled, but not as the word of YHWH itself as expressed in the Psalms. Even the prophets who claimed to be communicating the word of YHWH thought only to be speaking to their nation and the surrounding peoples of their time, not those across the globe who would read their words 3,000 years later.

So, when the apostle Paul penned the above to Timothy, he was making clear to his friend that the scriptures of the Old Testament were inspired by the intentions of the divine in spite of the authors' actual intentions. Scripture may have revealed truths about reality, but more importantly in this particular thought, scripture was useful for directing man to live a good life. Spinoza would have supported such a pragmatic view of scripture.

Spinoza's view was limited, however, and had to ignore portions of the very scripture that he said was useful, for the ancient scriptures, to which Paul lent credence in the above quote, also say:

> The unfolding of your words gives light:
> it imparts understanding to the simple. (Ps. 119:130)

Scripture claims itself to be, not only a helpful guide by which to live a good life, but also a lamp of understanding for those in the darkness of mortal ignorance.

There is a delicate balance for the theologian to strike. On one hand, he believes that the divine has spoken to him directly or through the scriptures, and through this communication, he feels that he has direct access to eternal truths; however, the careful theologian will also consider the sources of scripture and will have to temper his acceptance of the literal words with literary, allegorical, and historical interpretation. Perhaps in such interpretations he will gain access to greater truths. Or, perhaps he will dilute the divine words by perverting the original context of the revelation. The theologian, more so than the philosopher or scientist, believes that he has a higher authority to which he must answer. He must be extremely careful about what he says concerning scripture. In light of this fact, let us only say definitively about the divine what we can say without doubt. All other propositions must be labeled as they are: doubtful and less than divinely knowable truth at present.

Yet, although the theologian finds reason to be especially careful with his words, he also finds reason to be hopeful of the success of his search for truth.

For seekers of the water of life, God hath a mighty spring... No one who seeks His court will perish; neither will He disappoint the hopes of him who stands before His door. (Suhrawardi 162)

With this two-fold perspective, let us very carefully discuss what ancient theology teaches about the very foundations of the universe, which will give us a surprisingly comprehensive understanding of the theological view of light.

A CREATION STORY

In the beginning, God created the heavens and the earth. The earth was without form and void, and darkness was over the face of the deep. And the Spirit of God was hovering over the face of the waters. (Gen 1:1-2)

With these words, the Jewish and Christian scriptures begin, and if Muhammad is to be believed, Muslims also believe in these histories handed down through the ages, though not in the extant wording.

'In the beginning'. This phrase is difficult to understand; the greatest philosophers cannot quite conceptualize what this 'beginning' is. In the Newtonian worldview, all effects have causes, and cause requires time, for *before* there is not yet any effective cause and *after* the cause has already taken effect. The quantum physicist and the string theorist have an easier time understanding this concept of the 'beginning' than does the Newtonian, but that does not rid any thinker of contending with why the ultimate event happened, and how. Our modern scientific theories may lean on the necessity of a certain event happening due to particular physical laws and probabilities, but laws and probabilities are not a source of the prerequisite energies needed for creation. Where, and more importantly at present, *when* did these energies come from?

The Judeo-Christian reader (as well as the string theorist) must understand the first three words of Genesis as follows: in the creation of the heavens and the earth, we also have the creation of time. It is impossible to say what happened before creation. It is impossible to say what God was doing before creation. There is no *before creation* because there is no time apart from the physical creation. Time could not exist before there was something to measure it.

String theory agrees whole-heartedly with this concept of time and the lack thereof:

> But in the raw state, before the strings that make up the cosmic fabric engage in the orderly, coherent vibrational dance we are discussing, *there is no realization of space or time.* Even our language is too coarse to handle these ideas, for, in fact, there is even no notion of *before.* In a sense, it's as if individual strings are "shards" of space and time, and only when they appropriately undergo sympathetic vibrations do the conventional notions of space and time emerge. (Greene 378)

The child may have an easier time considering the infinity of space as opposed to the infinity of time, so many people gravitate to an understanding of the magnitude of the universe and try not to get caught up on the difficulties of eternity. However, the infinities of space are equally impossible to truly appreciate. There is no answer that satisfies the child about the infinite cosmos, and if the adult is honest, that dissatisfaction has probably followed her into her old age. Augustine suggests to his reader that this is an easy point to push back on the critic of religion:

> If they have an answer about the infinite spaces outside this world, if they can answer the question why God 'ceases from his work' in that infinity, then they can answer their own question about the infinity of time before the world, and why God was inactive then... Now if they assert that it is idle for men's imagination to conceive of infinite tracts of space, since there is no space beyond this world, then the reply is: it is idle for men to imagine previous ages of God's inactivity, since there is no time before the world began. (Augustine, City of God 435)

One useful way that scientists and philosophers understand time is by conceptualizing it as the means of measuring change in the world. In a completely static system with no change, there is no real concept of time. Time, in essence, is change. We can track the passing of the day because we observe the sun's circuit. We can track the passing of minutes as we watch our clock's whirling hands. We can track the passing of musical timing by the beats kept by the metronome. But there has to be change, be it cyclical or linear, in order for there to be an understanding of time and really for there to be any time at all. Time requires imperfect, changing, physical activity.

> [God] made all these things of which this mutable world consists, yet in a state of flux. Its mutability is apparent in the fact that passing time can be perceived and measured. (Augustine, Confessions 250)

> If we are right in finding the distinction between eternity and time in the fact that without motion and change there is not time, while in eternity there is no change, who can fail to see that there would have been no time, if there had been no creation to bring in movement and change, and that time depends on this motion and change, and is measured by longer or shorter intervals by which things that cannot happen simultaneously succeed one another? (Augustine, City of God 435)

Creation bears the existence of change and thus the existence of time. Eternity is existence without change, without time. So long as there is change, time exists and can be divided into smaller bits. When there is no change, there is no time, and a complete unity of experience abides. Muslims, Jews, and Christians understand their god to be eternal, unchanging, and unlimited by the constraints of time. He is "the Father of lights with whom there is no variation or shadow due to change" (James 1:17). God is eternal because he does not change. Man and his world are mortal because they do change. Before a mutable, physical world was created, time was not.

With a useable understanding of time, let us return to our first verse. "God created the heavens and the earth." It is important to understand the terms used in these short verses to avoid confusion with the rest of the creation story. Later, in verse 8 of this first chapter, we see God (in Hebrew, *Elohim*) giving the name 'Heaven' to an expanse that separated the earthly and cosmic 'waters' (the modern reader should understand the term 'waters' to refer to a physical formless void, an inexpressible 'deep'). The heaven of verse 8 is the sky: the firmament, the atmosphere, that which separates us from the star-spangled expanse above us.

In comparison with the heaven of heaven, even the heaven of our earth is earth. And it is not absurd to affirm that both of these vast physical systems are earth in relation to that heaven whose nature lies beyond knowledge, which belongs to the Lord, not to the sons of men. (Augustine, Confessions 246)

Heaven in verse 1 is not heaven in verse 8. There is a physical heaven (the cosmos) and a divine heaven (the spiritual realm). The physical heavens are a part of the physical creation and contained within the term 'earth' in verse 1. Verse 1 describes the first 'day' of God's creation, and verse 8 the second. Naturally, God did not create the same thing twice.

There is one easy interpretation of 'the heavens and the earth' of verse 1. To create space and time, God naturally has to create some physical and mutable thing in which time could be expressed. In this creation, God is simply defining a separation of the physical world and his perfect, higher reality. God creates the realm in which the physical will be housed, but he does not yet create physicality itself. God's blank canvas is 'earth'.

Verse 2 tells us that the earth was "without form and void" and that "darkness was over the face of the deep... the face of the waters." These waters are how the ancient writer of Genesis could conceive of physicality without any actual physical matter. The spirit of the divine hovered outside the physical un-physicality of the void, a concept without any possible mental construct. Void is as unimaginable as the absence of time. Human logic cannot contain this void, so Genesis instead calls it 'waters', some non-physical yet still physical thing. We cannot even properly call this void *space*, for what is space without something filling it? Where there is nothing but the possibility of the physical in a void realm, there is nothing, an incomprehensible darkness of complete absence.

Hinduism affirms this view of the origins of our universe in a portion of its oldest scriptures, the Rig Veda:

Then was not non-existent nor existent: there was no realm of air, no sky beyond it.
What covered in, and where? and what gave shelter? Was water there, unfathomed depth of water?
Death was not then, nor was there aught immortal: no sign was there, the day's and night's divider.
That One Thing, breathless, breathed by its own nature: apart from it was nothing whatsoever.

> Darkness there was: at first concealed in darkness this All was
>   indiscriminated chaos.
> All that existed then was void and formless: by the great power of
>   Warmth was born that Unit. (Rig 10:129.1-3)

Genesis 1 and Rig Veda 10 describe the same fundamental nothingness
of the beginning of our universe. All was void. But this void would not last.

> And God said, "Let there be light," and there was light. And God
> saw that the light was good. And God separated the light from the
> darkness. God called the light Day, and the darkness he called
> Night. And there was evening and there was morning, the first day.
> (Gen 1:3-5)

As we have already stated, in verse 2 the 'earth' is nothing but void,
absence, and darkness. Earth contains nothing and is nothing. In verse 3,
something profound happens. The word of God is released into the physical
realm in his first interaction with creation; his word is light, both the energy
of wave and the mass of particle. Elohim's utterance itself is the creative
impulse that impregnates the void with a thing, with the energy of light. As
the modern reader recognizes, light is a physical photon, and thus, God's word
is the first and, perhaps ultimately, the only physical thing to fill the physical
void of our universe. This light gives our universe substance. In a flash, light
is all that can properly be said to physically exist.

> Evidence supports the conclusion that what exists now in the
> universe is the result of creation that in religious belief began with
> the command "Let there be light," and as claimed by the physicists,
> it was the big bang. In either case, creation began with the
> application of energy, leading to the conclusion that matter is made
> of energy. That conclusion is supported by the equation $E = mc^2$
> where E = energy and m = mass and c is a multiplier to change from
> energy to mass units of measure. (Pelton 59)

So, light is released into the world. Energy and matter and space and
time are created simultaneously in the application of light. However, there
is still the existence of darkness. In spite of the light, Elohim maintains the
reality of the immaterial darkness by separating it from the real physicality of
the light. God realized (in that he made real, not came to the realization) that

light was good and removed it from the un-goodness of the dark. He gave each names, 'Day' and 'Night', though the sun, moon, and globe had not yet been created. These concepts of day and night must more closely align with light and darkness as opposed to the spinning of the globe toward and away from the sun. And so, the first day passes without a sun to set under a horizon that itself does not yet exist. Until the fourth day, there is no sun, moon, or stars. It can be difficult to defend the assertion that these 'days' of creation are 24-hour periods. Why would God constrain himself to the set limit of 24 hours before 24 hours was a concept and when the divine is not constrained by time anyway? But the question largely becomes immaterial due to lack of evidence or consequence one way or the other. There was physical light; this was enough for time to pass. And so time passed in these creative eras of indefinite length called 'Day', defined by the presence of the good light of Elohim.

On the second day of Judaic and Christian creation, Elohim separates the earth and heavens, as described above. These are the physical heavens that the psalmist describes:

> The heavens proclaim his righteousness,
>> and all the peoples see his glory. (Ps 97:6)

The physical cosmos is separated from the physical earth, a magnificent truth. The Qur'an goes so far as to claim that, "the creation of the heavens and earth is greater by far than the creation of mankind, though most people do not know it" (40:57); however, the balance of this book will be shown to argue against Muhammad's point of view. Regardless, now there are two entities, physical earth and physical heaven. Still there is morning and evening on the second day before there is a sun to rise or set. Time passes all the same.

On the third day, Elohim separates the wet, physical waters from dry earth and creates the continents in some form, calling the two 'Earth' and 'Seas'. Here we have a third understanding of *earth*. On the first day, it encompassed the entire physical realm. On the second day, it is everything contained within our globe. On the third day, earth is rock and soil and land. And on the land, God commanded that plant life would generate, and so it did, the first physical life we see is plant life. And there is morning and there is evening and there is no sun.

The fourth day sees Elohim populate the physical heavens with the stars, sun, and moon, creating a way for earthly beings to delineate time, though time itself had already been created as we have seen in the passage of the divine creative days. And the fourth day ends with, "And there was evening and there was morning, the fourth day" (Gen 1:19). This is the first moment where the statement that there was evening and morning actually makes sense to us earth-bound mortals. With the creation of sun and moon, we can properly understand the uniform passage of time and earthly days.

On the fifth day bird and marine life is created, and at the beginning of the sixth day, earth-bound life is created. The sixth day continues:

> Then God said, "Let us make man in our image, after our likeness. And let them have dominion over the fish of the sea and over the birds of the heavens and over the livestock and over all the earth and over every creeping thing that creeps on the earth.'
>
> So God created man in his own image,
>> in the image of God he created him;
>> male and female he created them. (Gen. 1:26-27)

The way that an individual understands these verses may come to define his understanding of all human life, so it is worthwhile to stop and analyze the text.

To be made in the image of the divine should be understood as a fact of great importance. Elohim is sharing his image, the way in which he is expressed in the physical world, with the final thing added to his creation: mankind. The question naturally becomes, "What is it that defines God's image?"

Though the divine character will come to be expressed in countless ways in the balance of scripture, the key should have already been stated in the 25 verses leading to this statement. God likely would not refer to his image before the reader has a reasonable opportunity to develop a basis for understanding that image. So we ask ourselves what, so far, defines who God is?

From Genesis 1:1, we know that Elohim is not confined by space and time. In fact, he created space and time at the instant of creation, at the beginning of time. We know that Elohim is not physical but removed from the physicality of the 'earth'. We know that he is spirit, at least in part, as he hovers over the void. We know that his utterance is power, creative energy, and the essence of light and the rest of creation.

We know that he creates all physicality with the words "Let there be...," but man he creates with the words "Let us make man in our own image." This is the first instance where we see Elohim refer to himself in the plural, either employing the *Royal We* or engaging in dialogue between his unified beings whose individual existences could not betray the absolute unity and perfect changelessness of eternity (the Hebrew *Elohim* itself is plural, though we rightly translate this divine being into one cohesive existence). We know that he creates man (mankind) to rule over the rest of his creation. We know that he creates man, both male and female, infusing them both with this divine character, God's own image, in the same utterance. We also know that starting in Genesis 2:4, God is referred to as *YHWH Elohim*, establishing his relationship with man by personalizing the generalized *Elohim* (used to describe any divinity of various traditions) to *YHWH* (used to refer to the Jewish understanding of a particular, personal, and interactive deity).

More than anything, at this point in Genesis, the primary aspect that comes to define our concept of Elohim and his image is his willful creative energy, the power with which he affected some sort of change in his perfect, changeless existence, purposefully creating the physical universe and placing man in the preeminent state above the rest of creation. This is the image of God. This is the image by which he created man. Man's essence is based on this character.

William Paul Young imagines a dialogue between one man and the triune God of Christianity in his book *The Shack*. God says to Mack, the main character, "As the crowning glory of creation, you were made in our image, unencumbered by structure and free to simply 'be' in relationship with me and one another" (Young 126). In Young's understanding, communion and relationship with God was the same as to be created in his image. Augustine gives his own opinion, referring to the mind of man, the instrument of divine reason:

[God] speaks to the highest of man's constituent elements, the element to which only God himself is superior. For man is rightly understood – or, if this passes understanding, is believed – to be made 'in the image of God'. And his nearness to God who is above him is certainly found in that part of man in which he rises superior to the lower parts of his nature, which he shares with the brute creation. And yet the mind of man, the natural seat of reason and understanding, is itself weakened by long-standing faults which darken it. It is too weak to cleave to that changeless light and to enjoy it; it is too weak even to endure that light. (Augustine, City of God 430)

For the philosopher, man's ability to reason is his highest element, and as man's greatest possession, our reason is the picture of the divine in us. For the theologian, it may be futile to try to understand the fullness of what is meant by 'in God's image'. As John the apostle says in his first letter recorded in the epistles of the New Testament, "Beloved, we are God's children now, and what we will be has not yet appeared; but we know that when he appears we shall be like him, because we shall see him as he is" (1 John 3:2). For John, the image of the divine cannot be fully expressed in man until man perceives the divine fully, a state that will be unattainable before the final day of judgment when God descends to earth.

If we are to understand, as this current work certainly contends, that the divine image in man is akin to the willful energy that defined Elohim's essence in the creation story, then man has significant stature in creation. The creative will in man has the power of independent agency that may or may not align with the original purposes of the divine will and yet somehow will not oppose the divine will. Man's will, if not properly employed, would be a force contradictory to God's good, an anticreative power that would wreak destruction on creation. As Elohim is the ultimate source of good and existence, man's will would be the ultimate source of evil and death. Elohim would have created the opportunity for evil, placing enormous responsibility in the hands of man, but the divine would not in any way be a perpetrator of those and all possible evils; man alone would bear the blame for misuse of the power he has been given.

The creation portion of the Biblical creation story has been sufficiently introduced at this point; however, the story continues to explain the source of man's separation from YHWH Elohim, a vital development. God was content to condescend into the limits of space and time in the garden of Eden, which he created for the man Adam and his wife Eve, to commune with his creation. Adam and Eve were to tend the garden, working the land and keeping the plant and animal life, walking with their YHWH Elohim daily in the coolness of Eden.

The tree of life was in the midst of the garden, and the tree of the
knowledge of good and evil... And the LORD God commanded the
man, saying, "You may surely eat of every tree of the garden, but of
the tree of the knowledge of good and evil you shall not eat, for in
the day that you eat of it you shall surely die." (Gen 2:9,16-17)

In Eden, YHWH Elohim created two specific trees, the tree of the
knowledge of good and evil and the tree of life. He commanded man not to
eat of the tree of the knowledge of good and evil, though no command kept
man from eating from the tree of life. However, man and his wife were
tempted by the evil will of Satan to partake of the first tree, and in partaking
found their condemnation.

It is important to realize that this is 'the tree of the knowledge of good
and evil', not 'the tree of knowledge'. YHWH does not wish to keep his
creation in complete ignorance but wishes that no condemnation or death fall
on man by means of his own misunderstanding of his distance from God's
sovereign perfection. In eating of the fruit of this tree, Adam and Eve
condemn themselves by gaining the understanding that they have reason to
be condemned. The act of eating the fruit is the realization of (in that Adam
and Eve both made real and came to understand) sin and death, the
misalignment of the human will with that of the divine. YHWH did not wish
for man to bring this condemnation upon himself, which was the reason for
the commandment to avoid the tree of the knowledge of good and evil.
YHWH desired that their existence might be unencumbered by the concerns
of a fading creation set up as a dichotomy to shining, divine goodness. YHWH
knew the character of his creation, but so long as they did not, so long as
Adam and Eve could not comprehend that their wills might contradict the
divine, they could have lived in perfect obedient communion with that
divinity eternally. But after Adam and Eve had eaten the fruit:

Then the LORD God said, "Behold, the man has become like one of
us in knowing good and evil. Now, lest he reach out his hand and
take also of the tree of life and eat, and live forever" – therefore the
LORD God sent him out from the garden of Eden. (Gen 3:22-23)

Before man willfully ignored YHWH (translated 'LORD' in the English
Standard Version of the Bible above) and his commandment to avoid the tree
of the knowledge of good and evil, Adam and Eve were free to eat of the tree
of life. YHWH did not forbid man to live eternally in communion with him

without sin. We assume that God would have preferred for man to do just that. However, after perverting the divine will and breaking his trust, YHWH would not allow Adam and Eve to attain the fullness of eternal life by their own willful means. Another way must be found by another will.

A number of generations passed from Adam and Eve's departure from the garden, during which time Cain killed Abel, as already described previously by Steinbeck. Cain's murder of Abel was only the beginning of a trail of evil deeds that began to define man's state on earth. YHWH saw that all men continued to pile wickedness upon wickedness by means of their distorted wills. "The LORD saw that the wickedness of man was great in the earth, and that every intention of the thoughts of his heart was only evil continually" (Gen. 6:5). 'The intention of the thoughts of his heart' can only be understood as man's wicked desire expressed through his active and evil will. This will had begun as YHWH Elohim's image in man, a great power and an awesome responsibility. Man misused it, generating evil, actualizing death, condemning the physical world, and bringing on himself the wrath of a good god, the very essence of the philosopher's goodness and light and virtue and truth.

GOD IS LIGHT

It is hardly possible to read any religious text or discuss someone's religious understanding without some kind of reference to the radiance and the good light of the divine. It seems that all humanity understands that somehow the character of their god is understood in terms of the warmth and goodness of light. Though every religious text will probably have some reference to it, the Judaic Pentateuch, Laws, and Prophets and the Christian Bible are replete with references to YHWH and God as light. The examples of this affinity for light seem to be on every page, but here we will try to pare back the examples to those that are most pertinent to the present discussion.

Generally, YHWH is poetically described as shining and brilliant, whether that be his clothing or his person or his general presence.

> Bless the LORD, O my soul!
> > O LORD my God, you are very great!
> You are clothed with splendor and majesty,
> > covering yourself with light as with a garment,
> > stretching out the heavens like a tent. (Ps. 104:1-2)

Out of Zion, the perfection of beauty,
  God shines forth. (Ps. 50:2)

YHWH is brilliant, and his light shines forth on those who see him directly, but also, importantly, it is understood that his light is universal, shining both on those who pay him devotion and those who do not even know his name. "He makes his sun rise on the evil and on the good, and sends rain on the just and on the unjust" (Matt. 5:45). If such a divinity existed, this is just what we would expect from the god that created all physicality in the beginning of time with his word, bringing forth his creative energies as light. All creation would be impregnated with his glorious character, and all would share alike in the luminosity of his world.

More than just these poetic descriptions of God's brilliance, often throughout the Bible, God expresses himself and interacts with people through real physical light. Moses was the first to witness the brightness of YHWH's fire, a flame that he would come to know with great familiarity.

He looked, and behold, the bush was burning, yet it was not consumed... And Moses hid his face, for he was afraid to look at God. (Ex. 3:2,6)

The burning bush gave off the light of a flame, but God's fire did not consume the bush as it rested upon it. YHWH's divine fire is noted for its light, not its destructive force, at least here. The divine is seen to interact with the world in some familiar and other dissimilar ways to that of mundane physicality. It is rightly called divine.

Just as God's presence and favor can shine upon his people and light their way, so too does YHWH express his disfavor in terms of light, but notably instead in the absence of light. After Moses had done numerous miracles in the sight of Pharaoh of Egypt in a vain attempt to convince the king to allow Moses' people, the Israelites, to travel into the desert to offer sacrifices to their god, YHWH begins to express his power in Egypt through the ten plagues. After eight plagues, Pharaoh still refused to allow the Israelites to venture into the desert.

Then the LORD said to Moses, "Stretch out your hand toward heaven, that there may be darkness over the land of Egypt, a darkness to be felt." So Moses stretched out his hand toward heaven, and there was pitch darkness in all the land of Egypt three days. They did not see one another, nor did anyone rise from his place for three days, but all the people of Israel had light where they lived. (Ex. 10:21-23)

Perhaps the absolute darkness described here is hyperbole, but there is not much interpretive help to make that claim. The words seem fairly literal. In Exodus 7, when YHWH turned the Nile and all Egyptian bodies of water into blood, the text says water in vessels of wood and stone would likewise turn to blood. These plagues are absolute and described in literal terms. Three mornings the sun rose and shed its light over the whole earth, but around the habitations of the Egyptians, none of the sun's rays had effect; however, in the slums of the Israelites, within the land of Egypt, the sun still shone. This is a supernatural proposition, and it is possible to understand it even more supernaturally.

The text does not make this clear, but it does say that the Egyptians did not rise from their homes, nor did they even see one another. It is not absurd to assume that the Egyptians' fires put off no light either, so that nothing at all could be seen, by the sun, moon, stars, or by fire. This is an interesting contrast to the fire Moses encountered at the burning bush. And in this supernatural darkness, the Egyptians had no option but to stay put.

Exodus shows God demonstrating his power in light and expressing his displeasure with darkness. Only a few chapters later, YHWH takes his place again as the guiding light of the Israelites.

And the LORD went before them by day in a pillar of cloud to lead them along the way, and by night in a pillar of fire to give them light, that they might travel by day and by night. (Ex. 13:21)

In this way YHWH led his people to Mount Horeb, or Mount Sinai, where the pillar of cloud and flame ascended the slopes and the presence of God rested on the mountain. And we are told, "Now the appearance of the glory of the LORD was like a devouring fire on top of the mountain in the sight of the people of Israel" (Ex. 24:17).

The fear that possessed Moses at the burning bush now comes to rest on the Israelites as the presence of YHWH smolders on the mountain, and in the face of the glory of God thundering above them, the Israelites plead with Moses to act as an intercessor for them.

Now when the people saw the thunder and the flashes of lightning and the sounds of the trumpet and the mountain smoking, the people were afraid and trembled, and they stood far off and said to Moses, "You speak to us, and we will listen; but do not let God speak to us, lest we die." Moses said to the people, "Do not fear, for God has come to test you, that the fear of him may be before you, that you may not sin." (Ex. 20:18-20)

The brilliance of YHWH's glory was masked behind the cloud of smoke that rose up from the mountain, and his people, those whom he had led out of Egypt with great miracles and for whom he parted the Red Sea so that they might escape Pharaoh's armies and whom he guided through the desert with a pillar of cloud in the day and fire at night, these people were so terrified by the radiance of his glory that they would not approach him. And Moses correctly tells his people that they should not wish for separation from this light but should allow their fear of its power to change them.

Moses, having encountered the goodness of the divine fire, recognized the power and the love behind it and continued in his role as the leader of the Israelites and the arbitrator between them and their god. He ascended the heights of the mountain to commune with the essence of the divine, and when he came down, the effects on his person were marked.

When Moses came down from Mount Sinai, with the two tablets of the testimony in his hand as he came down from the mountain, Moses did not know that the skin of his face shone because he had been talking with God. Aaron and all the people of Israel saw Moses, and behold, the skin of his face shone, and they were afraid to come near him. But Moses called to them, and Aaron and all the leaders of the congregation returned to him, and Moses talked with them. Afterward all the people of Israel came near, and he commanded them all that the LORD had spoken with him in Mount Sinai. And when Moses had finished speaking with them, he put a veil over his face.

Whenever Moses went in before the LORD to speak with him, he would remove the veil, until he came out. And when he came out and told the people of Israel what he was commanded, the people of Israel would see the face of Moses, that the skin of Moses' face was shining. And Moses would put the veil over his face again, until he went in to speak with him. (Ex. 34:29-35)

Thus the LORD used to speak to Moses face to face, as a man speaks to his friend. (Ex. 33:11)

We are told still more, that the presence of YHWH came off the mountain and again led the Israelites as it had, as a pillar through the wilderness. And when the people had obeyed his commandments and made the 'portable temple', called the *tabernacle*, God's presence would hover over the tabernacle whenever the people stopped to rest and remain in one place for a time. The cloud and fire would rest on and around the tabernacle until it was time for the Israelites to depart. Then the cloud would ascend and lead the way. "For the cloud of the LORD was on the tabernacle by day, and fire was in it by night, in the sight of all the house of Israel throughout all their journeys" (Ex. 40:38). God's presence and his miraculous fire were an ongoing miracle ever-present to the Israelites during their 40-year sojourn in the wilderness. His divine light never left them.

Moses tapped into the experience on the mountain when he explained YHWH's commandments to his people. The prophet draws a clear distinction between the radiance of their god and the lights that people see elsewhere in nature. Though the lights that populate the sky may be expressions of the glory of God in his creation, they are distinctly different from the actual radiance of the presence of YHWH Elohim. For this same reason, Genesis 1 employs the phrase 'greater and lesser lights' to describe the sun and moon. More commonplace names could have been used but would have then invoked the pagan deities of the time. The people must have understood all this, evidenced in the way that Moses commands them not to worship any other god.

Therefore watch yourselves very carefully. Since you saw no form on the day that the LORD spoke to you at Horeb out of the midst of the fire, beware lest you act corruptly by making a carved image for yourselves, in the form of any figure... And beware lest you raise your eyes to heaven, and when you see the sun and the moon and the stars, all the host of heaven, you be drawn away and bow down to them and serve them, things that the LORD your God has allotted to all the peoples under the whole heaven. (Deut. 4:15-16,19)

MARVELOUS LIGHT

YHWH walked in Eden with Adam and Eve. He spoke directly to Abraham, Isaac, and Jacob. His will was known to Joseph in Egypt. However, until God condescended to commune with the people of Israel in the desert, no mortal described his physical presence on earth. YHWH was clearly shown to be radiant with light, imposing that light upon the person of Moses, and masking the light behind a dark cloud of smoke for the benefit of his people who were terrified by direct interaction with him.

Just as the divine is literally described this way in his physical expression on earth, so too are we told by the prophets that God's person and presence shine in heaven. Ezekiel describes the divine throne in the prophetic visions he experienced by the Chebar canal:

> And above the expanse over their heads there was a likeness of a throne, in appearance like sapphire; and seated above the likeness of a throne was a likeness with a human appearance. And upward from what had the appearance of his waist I saw as it were gleaming metal, like the appearance of fire enclosed all around. And downward from what had the appearance of his waist I saw as it were the appearance of fire, and there was a brightness around him. Like the appearance of the bow that is in the cloud on the day of rain, so was the appearance of the brightness all around.
> Such was the appearance of the likeness of the glory of the LORD. (Ez. 1:26-28)

It is almost comical how careful Ezekiel is to express his impression without making the attempt to describe the glory of YHWH exactly. 'The appearance' and 'the likeness' are ways in which Ezekiel makes it clear that it was impossible to describe exactly what he saw. And even in describing the appearance, he seems to grasp for words. He refers to sapphire and metal and brightness and rainbows. The prophet does not even have a direct earthly description of the divine upper body and has to say that it was like 'fire enclosed all around'. There is probably no book of the Bible that has stronger imagery and fanciful description than Ezekiel, and yet the words here are clearly lacking. YHWH's glory cannot be described. But if we attempted to describe it, we would have to think of the most resplendent and shining things on earth to encapsulate the radiance of the divine presence.

Beings in heaven, other than he who resided above the throne, share similar qualities. Ezekiel describes angelic beings:

As for the likeness of the living creatures, their appearance was like burning coals of fire, like the appearance of torches moving to and fro among the living creatures. And the fire was bright, and out of the fire went forth lightning. And the living creatures darted to and fro, like the appearance of a flash of lightning. (Ez. 1:13-14)

So too is the glorified Christ described in the New Testament corollary to Ezekiel's prophecy, Revelation:

The hairs of his head were white, like white wool, like snow. His eyes were like a flame of fire, his feet were like burnished bronze, refined in a furnace, and his voice was like the roar of many waters. In his right hand he held seven stars, from his mouth came a sharp two-edged sword, and his face was like the sun shining in full strength. (Rev. 1:14-16)

Furthermore, in describing the new heaven and the new earth which will appear after God's redemption of his physical creation, we see the person and presence of the divine shine forth in even more glorious ways:

Then I saw a new heaven and a new earth, for the first heaven and the first earth had passed away, and the sea was no more. And I saw the holy city, new Jerusalem, coming down out of heaven from God, prepared as a bride adorned for her husband... And I saw no temple in the city, for its temple is the Lord God the Almighty and the Lamb. And the city has no need of sun or moon to shine on it, for the glory of God gives it light, and its lamp is the Lamb. By its light will the nations walk, and the kings of the earth will bring their glory into it, and its gates will never be shut by day – and there will be no night there. (Rev. 21: 1-2, 22-25)

If John, the writer of Revelation, was describing the same divinity Moses encountered, we must understand why Moses ordered the people to refrain from worshipping the celestial lights. In Revelation, we see that God himself will be the ultimate and only source of light, creating one eternal day with his omnipresence. Even if still extant, what role would the sun and moon and stars play when they cannot outshine the light that will always shine in the new Jerusalem? Isaiah witnesses the same phenomenon in his revelations:

The sun shall be no more
    your light by day,
nor for brightness shall the moon
    give you light;
but the LORD will be your everlasting light,
    and your God will be your glory.
Your sun shall no more go down,
    nor your moon withdraw itself;
for the LORD will be your everlasting light,
    and your days of mourning shall be ended. (Is. 60:19-20)

During their wanderings in the wilderness, the Israelites experienced the physical glory of YHWH. In the revelations of the prophets, we further see the radiance of God universally shining forth after his creation has been redeemed from the darkness of sin. God the Father is resplendent in his glory throughout the Old and New Testaments of the Bible; however, Christ his Son and other mortals and earthly beings also shine the glow of the divine on earth throughout the stories of the New Testament.

CHRIST THE MESSIAH

Ancient Judaic interpretation of the prophecies recorded in the scriptures accounted for a messiah who would come to save YHWH's people from the ravages of the hostile world in which they lived. This messiah was often seen as a political or militaristic figure who would reclaim the territories claimed by Israel as the land of the nation's forefathers, Abraham, Isaac, and Jacob. Jerusalem would thus become the capital of the world through which YHWH would reign on earth in some capacity.

The Christian perspective of those same prophecies reinterprets this messianic figure as Christ Jesus, who came to save God's true children within and without the nation of Israel. This messiah, in his life and actions on earth, paved the way for God's people to be redeemed through the divine plan of reconciliation between earth and heaven, ushering a new age of complete communion with God not experienced since Adam and Eve's expulsion from the garden. Without this messiah, the human will and the divine will could not coexist eternally.

The Jewish messiah has yet to appear. However, in the story recounting this Christ through whom Christians bear their name, we see a continuation of the poetic and literal expressions of the divine light on earth.

The birth of Jesus was demarcated by two lights appearing in the sky. The wise men, familiar in the Christmas story, lived far east of Israel and saw one of those lights. We are made to understand that these men were familiar with the Judaic scriptures and prophecies, possibly due to Israel's exile in Babylon and Assyria, during which time Jewish tradition would have been communicated through the likes of Daniel to the sages of the east. It is said that the wise men observed a new star rise in the sky, a star they took to represent the birth of the Jewish messiah who would redeem the earth. The magi thus prepared for the long journey, following the star to where it would direct them, to this messiah.

The learning, preparation, and determination of the wise men is contrasted against the happy fortune of the uneducated shepherds who were out watching their flock on the night of Christ's birth. Instead of descrying a faint and distant light, we are told that the shepherds were dazzled by the divine:

> And in the same region there were shepherds out in the field, keeping watch over their flock by night. And an angel of the Lord appeared to them, and the glory of the Lord shone around them, and they were filled with great fear. (Luke 2:8-9)

In the Christian understanding of these events, we are to believe that the Son of God, divinity himself, in the form of a baby, entered into the world. Insofar as this could be true, it of course makes sense that the condescension of the divine in our world would again be expressed through light, as it had been in the Old Testament.

Jesus' life, as recorded in the New Testament, was hardly typical, but his presence is not as often described as being light-like, as YHWH's was in the Old Testament. Jesus was a man and hardly ever bore the physical mark of God his Father like Moses' shining face, but there are two notable instances of his divinity being expressed through light.

The first is when he ascended a mountain with his closest friends and disciples. There were only three men with him to witness this event, unlike the entire nation of Israel at Mount Horeb, so the skeptic may have a harder time accepting this story from the New Testament though it is far more recently recorded than Moses' account.

> And after six days Jesus took with him Peter and James, and John his brother, and led them up a high mountain by themselves. And he was transfigured before them, and his face shone like the sun, and his clothes became white as light. And behold, there appeared to them Moses and Elijah, talking with him. And Peter said to Jesus, "Lord, it is good that we are here. If you wish, I will make three tents here, one for you and one for Moses and one for Elijah." He was still speaking when, behold, a bright cloud overshadowed them, and a voice from the cloud said, "This is my beloved Son, with whom I am well pleased; listen to him." When the disciples heard this, they fell on their faces and were terrified. (Matt. 17:1-6)

Peter, overwhelmed by the scene he was witnessing, began to ramble and was stopped short by a voice that emanated from the presence of this spectacle. The whole description of this event aligns so closely with the other instances of divine brilliance that we have already found throughout the Jewish and Christian scriptures. Jesus, in this sense, is at least compared to the prophecies of the Old Testament, if not directly with God the Father.

Not long after this event, Jesus would be led to his death by the anger and false testimonies of the religious leaders in Jerusalem. The statements he continued to make throughout the four gospels were either the most extraordinary claims ever made or the highest blasphemy. In reading the accounts of his disciples, it is not hard to appreciate why the Pharisees were so consternated by the person of Jesus. He said outrageous things against their understanding of scripture.

When they finally succeeded in acquiring permission to have him crucified on a Roman cross, the Jewish leaders were glad to have Jesus be a concern of the past. However, on the cross, from noon to three o'clock in the afternoon, the Bible tells us that something extraordinary happened to match the claims Jesus had made. "Now from the sixth hour there was darkness over all the land until the ninth hour" (Matt. 27:45). 'Land', in this verse, may also be appropriately translated as 'earth'. As exclusive as Jesus' transfiguration was on the mountaintop with only his three closest disciples, the darkness that overtook the land at his death may have been a global event, surpassing even the grandeur of YHWH's presence on the mountain. In the death of the divine messiah, God's presence was, in a sense, removed from earth, prompting Jesus to feel forsaken, having been separated from the light of the Father's glory. God's presence is felt in unbearable shining like that of the sun, and his absence is felt in the weight of darkness experienced in the Egyptian plague and as Jesus hung on the cross.

But the darkness would not last. The light was to return. On the morning of Jesus' supposed resurrection:

> Now after the Sabbath, toward the dawn of the first day of the week, Mary Magdalene and the other Mary went to see the tomb. And behold, there was a great earthquake, for an angel of the Lord descended from heaven and came and rolled back the stone and sat on it. His appearance was like lightning, and his clothing white as snow. (Matt. 28:1-3)

The resurrection of Christ on earth is said to have been ushered in by a new dawn, and as the rays of the sun crested the horizon, a radiant angel of God came to meet the two Mary's. God's light returned.

The followers of Christ, after his death and resurrection, encountered continued expressions of the divine through light. When Peter was arrested and sent to jail, "an angel of the Lord stood next to him, and a light shone in the cell" (Acts 12:7), from which he was subsequently released by the angel. Likewise, Saul, later the Apostle Paul, a great persecutor of the early church, encountered God's presence when he was travelling to kill Christians and destroy churches north of Jerusalem:

> Now as he went on his way, he approached Damascus, and suddenly a light from heaven shone around him. And falling to the ground he heard a voice saying to him, "Saul, Saul, why are you persecuting me?" And he said, "Who are you, Lord?" And he said, "I am Jesus, whom you are persecuting. But rise and enter the city, and you will be told what you are to do." The men who were traveling with him stood speechless, hearing the voice but seeing no one. Saul rose from the ground, and although his eyes were opened, he saw nothing. So they led him by the hand and brought him into Damascus. And for three days he was without sight, and neither ate nor drank. (Acts 9:3-9)

We again see the glory of Jesus expressed in the same terms by which YHWH is recognized. If the Christian claim that Jesus is the Son of God, somehow God himself, is to be worth consideration, it would be necessary for Jesus to share his father's characteristics. The New Testament descriptions are not lacking. The inquisitive reader ought to go on to explore how else this messiah is described through the scriptures.

> The people who walked in darkness
>> have seen a great light;
> those who dwelt in a land of deep darkness,
>> on them a light has shone. (Is. 9:2)

Isaiah's description of the messiah is clear. He sees this figure as being a great source of radiance in a world in which sin and turmoil has obscured the view of the divine light. The messiah would be an earthly source of light, illuminated directly by YHWH's brilliance and reflecting his righteousness. The authors of the New Testament picked up on this theme and developed it in detail.

> The true light, which gives light to everyone, was coming into the world. He was in the world, and the world was made through him, yet the world did not know him. He came to his own, and his own people did not receive him. But to all who did receive him, who believed in his name, he gave the right to become children of God, who were born, not of blood nor of the will of the flesh nor of the will of man, but of God. (John 1:9-14)

There is far too much in these six verses to unpack fully at present, as they touch on many topics this book concerns; however, we here can recognize that the messiah of YHWH, sent by God himself, would have been of God and in some way would have shared in the character of YHWH, the creator of the physical universe. As God is light, so the messiah must be light-like.

In the Christian understanding of the messiah, Jesus claimed to be God's own son, of God himself, somehow a manifestation of the same being. The trinity, as believed by Christians, is a complex concept that will be treated in a little more depth later, but for now we can appreciate the thought that Jesus was divine, sharing God's character in a direct relationship.

He is the radiance of the glory of God and the exact imprint of his nature, and he upholds the universe by the word of his power. (Heb. 1:3)

He is the image of the invisible God, the first born of all creation. For by him all things were created, in heaven and on earth, visible and invisible... all things were created through him and for him. And he is before all things, and in him all things hold together. (Col. 1:15-17)

Again, it seems that there is simply too much contained in these short verses to treat without treating the whole topic fully, which will only be accomplished by finishing the entirety of this work, but Jesus, in the Christian view, embodies all of the same character of God. As God's Word, Jesus is seen as the declarative and creative sentences breathed out at the beginning of time. These are magnificent claims by the early Christians, but they were not any bolder than the statements that Jesus Christ made himself: "I am the light of the world. Whoever follows me will not walk in darkness, but will have the light of life" (John 8:12).

Although most of the examples of this kind of imagery being employed to describe the messiah is found in the New Testament, in his testimony in front of King Agrippa in Caesarea, Paul the Apostle, a man of great learning in the tradition of the Pharisees, of supreme knowledge of the Judaic texts, claimed:

To this day I have had the help that comes from God, and so I stand here testifying both to small and great, saying nothing but what the prophets and Moses said would come to pass: that the Christ must suffer and that, by being the first to rise from the dead, he would proclaim light both to our people and to the Gentiles. (Acts 26:22-23)

Paul, in his legal defense of Christianity, drew heavily on the Jewish Old Testament. There were those present who heard these words who thought Paul to be insane, but Agrippa, a man of excellent learning like Paul, is recorded saying, "In a short time would you persuade me to be a Christian?" (Acts 26:28). Paul's interpretation of the ancient scriptures was compelling enough for King Agrippa to concede that he might be convinced of the same. The king may have feared as much, not wanting to find himself in the middle of the controversies of this Jesus. To many Jews in those early days, Paul's arguments were so cogent that they did believe and formed the beginnings of what quickly became a global revolution in religion. To them, it was possible and very likely that the light that Paul and Jesus described was the same light described in the Jewish scriptures.

DIVINE LIGHT OF MAN

Even if the reader of the Bible disregards the literal representations of God as light, he will encounter time and time again ways in which the writers suggest that the goodness, righteousness, and salvation of man should be understood as being like light. In the Old Testament, after the effort of Queen Esther affected the salvation of the Jews in Persia, it was said that, "The Jews had light and gladness and joy and honor" (Esth. 8:16). The goodness experienced by the displaced nation is light, and the goodness of YHWH is the same. "Your word is a lamp to my feet and a light to my path" (Ps. 119:105). The Psalmist agrees that the influence of YHWH in his life is this good light. Even the Persians who took the Jews into exile are said to perceive the same qualities of divine goodness and wisdom. In talking with the king of Persia, his advisors commented:

> There is a man in your kingdom in whom is the spirit of the holy gods. In the days of your father, light and understanding and wisdom like the wisdom of the gods were found in him, and King Nebuchadnezzar, your father – your father the king – made him chief of the magicians, enchanters, Chaldeans, and astrologers, because an excellent spirit, knowledge, and understanding to interpret dreams, explain riddles, and solve problems were found in this Daniel. (Dan. 5:11-12)

Daniel's wisdom and understanding are referred to as light from the gods. The Persians recognized a divine brilliance in him and described him as such. YHWH's grace of bestowing insight upon his people is one form of his light in this world, but still, at any given time, the totality of the divine understanding is not perceived, and the devout must wait for future revelation and understanding to appreciate the perfect light of the divine: "Therefore do not pronounce judgement before the time, before the Lord comes, who will bring to light the things now hidden in darkness and will disclose the purposes of the heart" (1 Cor. 4:5).

The New Testament authors were constantly very careful in their language, vacillating between glorying in the good luminosity of God's arrival on earth while also waiting patiently for that goodness to come in its fullness. Though the messiah had come for the believers and affected their salvation, it was not quite right for them to suppose their salvation was complete in their time on earth:

> And we have the prophetic word more fully confirmed, to which you will do well to pay attention as to a lamp shining in a dark place, until the day dawns and the morning star rises in your hearts. (2 Peter 1:19)

The effects of the divine make certain truths real in our time, yet they will only be fully affected at the end of time, through the lens of eternity. Until then, we live in a superimposed state, waiting for the future to realize our present. Regardless of the timing, however, both Jewish and Christian authors agree that their salvation will come to earth in a blaze of light, as seen in the New and Old Testament quotes below, respectively:

> For my eyes have seen your salvation
> > that you have prepared in the presence of all peoples,
> a light for revelation to the Gentiles,
> > and for glory to your people Israel. (Luke 2:30-32)

> Arise, shine, for your light has come,
> > and the glory of the LORD has risen upon you.
> For behold, darkness shall cover the earth,
> > and thick darkness the peoples:
> but the LORD will arise upon you,
> > and his glory will be seen upon you.
> And the nations shall come to your light,
> > and kings to the brightness of your rising...
> Then you shall see and be radiant;
> > your heart shall thrill and exult. (Is. 60:1-3,5)

YHWH's salvation is coming, whether in part or in whole, and his people will witness the divine glory. In turn they will reflect God's glory to the rest of his world, populated by non-believers who will be drawn to this good light. Only at the time of this revelation will the world see the divine light, and until then we are forced to consider, "Are your wonders known in the darkness, or your righteousness in the land of forgetfulness?" (Ps. 88:12).

SINFUL DARKNESS OF MAN

Is not the day of the LORD darkness, and not light,
and gloom with no brightness in it? (Amos 5:20)

Hold on. We are told time and time again through the scriptures that YHWH is light and that his presence sheds light and goodness upon all the earth. Where is Amos getting the idea that on the day YHWH comes there will be an evil gloom and darkness?

Here we are introduced to one of the great dichotomies of scripture. God is light and goodness to those who believe in him and obey him with a fearful love, but conversely, YHWH will be perceived as a great darkness to those on whom he must exercise his judgement. God is good and cannot abide evil. We are made to understand that the divinity of our world will affect the removal of sin from its redeemed creation, allowing for nothing that will corrupt the perfection of the new heaven and earth. Sin would be such an evil force and must be exiled.

In this way, though the scriptures often tell us about YHWH's glory and light and radiance, we are also told that this goodness is veiled from us so that the complete glory of God would not overwhelm us sin-laden mortals. When Solomon finished the temple of YHWH, establishing a set abode, as opposed to the tabernacle used by the Israelites in the wilderness, we are told that God's glory descended on the temple like it had done to the tabernacle.

And when the priests came out of the Holy Place, a cloud filled the house of the LORD, so that the priests could not stand to minister because of the cloud, for the glory of the LORD filled the house of the LORD. Then Solomon said, "The LORD has said that he would dwell in thick darkness." (1 Kings 8:10-12)

This seems contradictory to how we understand the presence of God, but it is not unprecedented in the scripture. Though Moses descended from Mount Horeb with his face shining due to YHWH's presence, the mountain was shrouded by a cloud of smoke and thick darkness. Likewise, the glory of YHWH was obscured in the tabernacle. The sinfulness of the world can only have glimpses of his glory. Man cannot perceive it in its totality. Moses himself, we are told, only caught sight of the true nature of YHWH in passing. The intentional obscuration of God's full glory speaks far more to the relationship between God and man than to God's own character.

Man is shadow walking in a world of darkness, and it is God's light that offers him the opportunity to enjoy the security of the day. In one of the most well-known verses of all of the Jewish and Christian scriptures, King David says, "Even though I walk through the valley of the shadow of death, I will fear no evil, for you are with me" (Ps. 23:4). The world is in darkness. Man's sinful will affects that darkness. God is light. He chases away the shadow of sin.

God bears so much light and illuminates all he perceives, so it is impossible to say that the creation is true darkness to him. "Even the darkness is not dark to you; the night is bright as the day, for darkness is as light with you" (Ps. 139:12). So too, we are allowed to imagine that when the brilliance of the divine encounters the darkness of man, man himself seems less real. Our darkness is the privation of light, not a thing in itself. Damned men can be pictured as follows when they encounter the all-pervasive light of God: "Now that they were in the light, they were transparent – fully transparent... They were in fact ghosts: man-shaped stains on the brightness of that air" (Lewis, The Great Divorce 321).

THE ILLUMINATION OF MAN

Sinful man, all humanity, is shadowy darkness. Mortal man lacks the thick, shining reality of the divine. But we are made to believe that man has access to gain God's real light.

Since we have such a hope, we are very bold, not like Moses, who would put a veil over his face so that the Israelites might not gaze at the outcome of what was being brought to an end. But their minds were hardened. For to this day, when they read the old covenant, that same veil remains unlifted, because only through Christ is it taken away. Yes, to this day whenever Moses is read a veil lies over their hearts. But when one turns to the Lord, the veil is removed. Now the Lord is the Spirit, and where the Spirit of the Lord is, there is freedom. And we all, with unveiled face, beholding the glory of the Lord, are being transformed into the same image from one degree of glory to another. (2 Cor. 3:12-18)

While you have the light, believe in the light, that you may become sons of light. (John 12:36)

Those who look to him are radiant,
     and their faces shall never be ashamed. (Ps. 34:5)

In the Judeo-Christian understanding of the divine, the world has a god who desires to share the light and reality of his existence with the rest of his creation. He wants unity between heaven and earth, and he will work to affect that reunion. Christian authors seem to have a bit more to say on this point because they believe to have seen the beginnings of the process. The early Christian thinkers who penned the text of the New Testament begged their fellow believers to accept the glorious light of God.

For at one time you were darkness, but now you are light in the Lord. Walk as children of light (for the fruit of light is found in all that is good and right and true). (Eph. 5:8-9)

Through the redemptive sacrifice, grace, and mercy of the Christ Messiah, Christians believe that they might bear the character of God and ought to represent his radiance in the world. As Jesus describes:

You are the light of the world. A city set on a hill cannot be hidden. Nor do people light a lamp and put it under a basket, but on a stand, and it gives light to all in the house. In the same way, let your light shine before others, so that they may see your good works and give glory to your Father who is in heaven. (Matt. 5:14-16)

This language echoes the Old Testament prophetic visions of the New Jerusalem, which would represent the presence of the divine in the redeemed earth where the kings of men would look and come to pay their respect to the king of kings. In the New Testament view, as expressed above, women and men who personally experience God's salvation are that allegorical city, shining through the earth and showing our world the true king. Redeemed man becomes the focal point of the new prophetic imagery. What was before seen as the literal nation of Israel is translated into all believers who are adopted into the family of God as his children, not only natural born Israelites. The individual gains greater importance through the work of the Christian messiah. The soul and will of man are reestablished as a divine tool in God's creation.

> While in other sciences the instruments you use are things external to yourself (things like microscopes and telescopes), the instrument through which you see God is your whole self. And if a man's self is not kept clean and bright, his glimpse of God will be blurred. (Lewis, Mere Christianity 90)

> The eye is the lamp of the body. So, if your eye is healthy, your whole body will be full of light, but if your eye is bad, your whole body will be full of darkness. If then the light in you is darkness, how great is the darkness! (Matt. 6:22-23)

The above quotes are rightly interpreted metaphorically, but there are those who describe this transformation literally. William Paul Young's *The Shack* has already been quoted; however, the emotional and thematic climax of the novel is a great example of how some like to imagine what the literal light of God might look like in humans. Mack, the main character, is led into a wide-open field at night which is illuminated by a supernatural happening. Human forms swarm the field and radiate a glow that lights the whole scene. "There were no candles – they themselves were light. And within their own radiance, each was dressed in a distinctive garb" (Young 213). The light of God shone through the assembly, lighting the field, yet the individual personalities were still expressed through the particular colors of each person's glow. Even more, the personality of Mack's father, who was abusive while he lived, so abusive that Mack purposefully poisoned and killed him, could not help but burst through the general glow of the field. The long-lost love of their broken relationship sparked and shot out lightning bolts that arched between the two men until they came together in a glorious reunion. It is a beautiful picture of the redeemed state of man, though of course, it is only an imaginary depiction.

There are those, however, who believe this type of description to be a real explanation of man's energy and presence. Her experiences may be a little far-fetched and difficult to believe and may be chalked up to hallucination, but Jill Bolte Taylor described the same type of phenomenon, which she literally experienced, in her book *My Stroke of Insight*. As a neuroanatomist, Taylor had a unique perspective when she experienced her own stroke. Few people could recognize what was happening to them as it was happening like Taylor did. However, importantly for us at this point, Taylor claimed that she experienced the light of individuals, like Mack did, while she was still in the hospital after her stroke.

> I experienced people as concentrated packages of energy. Doctors and nurses were massive conglomerations of powerful beams of energy that came and went. (Taylor 74)

Taylor's explanation of this phenomenon is uniquely colored by her own personal beliefs that align with eastern religions; however, this is an example of man's light being literally expressed in our world in our time. Believe what you can about Taylor's experience, but one can hardly ignore the correlations that we see between her story and all that this book has treated to this point. The reader should not be surprised to see these correlations of light between science, philosophy, and western religion extend to the religions and philosophies of the east. These ideas seem to be truly universal.

RISING IN THE EAST

> Know that my brilliance,
> flaming in the sun,
> in the moon, and in fire,
> illumines this whole universe. (Gita 15:12)

In the Qur'an of Islam, the Bhagavad Gita and Vedas of Hinduism, and the Dhammapada of Buddhism, we see references to the divine and its light which are remarkably akin to the divinity of Judaism and Christianity. It is as if individuals within each faith had a view of the same general divinity, and though they interpreted that divinity differently, very differently, they could hardly help describing it in the same way.

> If the light of a thousand suns
> were to rise in the sky at once,
> it would be like the light
> of that great spirit.

Arjuna saw the universe
in its many ways and parts,
standing as one in the body
of the god of gods. (Gita 11:12-13)

The divine is light; it illumines the universe and shares its light with man. The highest divinity of most religions is believed to have created light and maintains it in our universe:

Thou, making light where no light was, and form, O men: where
form was not,
Wast born together with the Dawns. (Rig 1:6.3)

In Soma's ecstasy Indra spread the firmament and realms of light...
By Indra were the luminous realms of heaven established and
secured,
Firm and immovable from their place. (Rig 8:14.7,9)

In the above Hindu text, the creation of the physical world is related to the creation of light and the luminous realm of the heavens. Light is a key feature of form, by which creation attains its physical expression. Light comes into the world at the same time as physicality, and the physical seems to be an expression of this creative light. In this way, the Rig Veda echoes the creation story from Genesis.

In the Genesis account, we saw that YHWH placed the heavenly lights in the sky in part for man to be able to keep track of the passing time. Likewise, in the Qur'an we are told,

It is He who made the sun a shining radiance and the moon a light,
determining phases for it so that you might know the number of
years and how to calculate time. (Qur'an 10:5)

And just as YHWH created the universe through light but is also described in terms of light himself, so too is the divine described outside of the Judeo-Christian tradition:

God is the Light of the heavens and earth. His Light is like this:
there is a niche, and in it a lamp, the lamp inside a glass, a glass like

a glittering star, fueled from a blessed olive tree from neither east nor west, whose oil almost gives light even when no fire touches it – light upon light – God guides whoever He will to his light; God draws such comparisons for people. (Qur'an 24:35)

The shining character of the divine bringing wisdom, happiness, and joy transcends religious division:

Shine on us with thy radiant light, O Ushas, Daughter of the Sky,
Bringing to us great store of high felicity, and beaming on our
    solemn rights. (Rig 1:1.48)

God expresses his character in the world, and following his example, the devotion of believers reemphasizes the glorious light of the divine, perhaps as another useful expression of that light to the unbeliever:

To thee, dispeller of the night, O Agni, day by day with prayer
Bringing thee reverence, we come.
Ruler of Sacrifices, guard of Law eternal, radiant One. (Rig 1:1.7-8)

But the luminous character of the divine is not quite perceived even by those who dedicate their devotion to him. There will be a distance, literal or figurative, that will not be totally overcome in our time, though the effect of the light will be felt and enjoyed:

He is our father, our begetter, the ordainer,
Who begot us from being unto being,
Who alone assigneth their names to the Gods,
Him other beings approach for knowledge.
Wealth they won by offering to him
The seers of old like singers in abundance,
They who fashioned these beings illumined and unillumined
In the expanse of space.
Ye shall not find him who produced this world;
Another thing shall be betwixt you;
Enveloped in mist and with stammering
The singers of hymns move enjoying life. (Black Yajur 4:6:2.C-E)

THEOLOGY OF LIGHT

The divine is a guiding source of light for those who believe, a light that will dispel the darkness, a darkness that makes it difficult for humanity to move through this physical realm with divine confidence:

Conceal the horrid darkness, drive far from us each devouring fiend.
Create the light for which we long. (Rig 1:86.10)

God is the ally of those who believe: He brings them out of the depths of darkness and into the light. (Qur'an 2:257)

In part, the darkness that envelops the physical realm is brought upon man by man himself. The sin and disbelief of man contributes to and is perhaps the source of the darkness:

They try to extinguish God's light with their mouths, but God insists on bringing His light to its fulness. (Qur'an 9:32)

May his light chase our sins away. (Rig 1:97.1)

All that is good in man is perceived as coming from the divine. In more extreme theologies of determinism, man cannot even make his own decisions, and the divine affects the light that shines through the mortal. Regardless of the particulars, the good character of man is seen as coming from God:

That which is wisdom, intellect, and firmness, immortal light which
creatures have within them,
That without which men do no single action, may that, my mind,
be moved by right intention.
Whereby, immortal, all is comprehended, the world which is, and
what shall be hereafter...
Controlling men, as, with the reins that guide them, a skillful
charioteer drives fleet-foot horses. (White Yajur 34:3-4,6)

Regardless of how he comes to share in the divine light, somehow man has access to become illumined himself. In Buddhism, by practicing one's duties, a man can attain this light:

The sun is ablaze by day.
The moon shines at night.

The warrior is ablaze arrayed.
The superior one is ablaze meditating.
And every day and every night,
the awakened one is ablaze with splendor. (Dhammapada 26:387)

After a mortal has encountered the light of the divine, he might have access to the immortality of that light. It is possible that the devout will attain this radiance in this life or the next:

A seeker will master this earth,
this world of death and radiant beings. (Dhammapada 4:45)

Men who know the infinite spirit
reach its infinity if they die
in fire, light, day, bright lunar night,
the sun's six-month northward course. (Gita 8:24)

A disciple of the fully awakened one,
by means of insightful knowledge, shines. (Dhammapada 5:59)

And finally, as the followers of the divine die and enter into the paradise of the hereafter, they will be literally transformed into some embodiment of light shared with them from the character of the divine:

With their lights streaming out ahead of them and to their right,
they will say, 'Lord, perfect our lights for us and forgive us: You
have power over everything.' (Qur'an 66:8)

Almost all theological points concerning light in the Jewish Old Testament and the Christian New Testament are also described in the faith traditions outside of the Judeo-Christian conception. It seems that man's understanding of the divine transcends geographical and religious boundaries, uniting humanity through this divine light.

We would do well to explore in much more detail the implications of these beliefs and how they tie in with our current philosophical understanding. When we continue to find correlations between these disciplines, it should not surprise us in the least that we also find similar characteristics in the natural world of science.

Now, having treated the science, philosophy, and theology of light separately, in the balance of this book we will consider particular aspects of physical light and how its characteristics are expressed in comparable ways again and again in human life, giving structure and legitimacy to all ways in which we study our reality. The facts of science, as we understand them, support and develop deeper truths in wonderfully rich and astounding breadth.

The world is a way. We will soon find that no matter how it is studied, it expresses the same qualities time and time again.

MARVELOUS LIGHT

DUALITY OF REALITY

Experiments show that Einstein's particles of light are quite different from Newton's. Somehow photons – although they are particles – embody wave-like features of light as well. The fact that the energy of these particles is determined by a wave-like feature – frequency – is the first clue that a strange union is occurring. But the photoelectric effect and the double-slit experiment really bring the lesson home. The photoelectric effect shows that light has particle properties. The double slit experiment shows that light manifests the interference properties of waves. Together they show that light has *both wave-like and particle-like properties*. The microscopic world demands that we shed our intuition that something is either a wave or a particle and embrace the possibility that it is *both*. It is here that [Richard] Feynman's pronouncement that "nobody understands quantum mechanics" comes to the fore. We can utter words such as "wave-particle duality." We can translate these words into a mathematical formalism that describes real world experiments with amazing accuracy. But it is extremely hard to understand at a deep, intuitive level this dazzling feature of the microscopic world. (Greene 102-3)

Modern science tells us clearly that submicroscopic, subatomic entities must be treated as both waves and particles; however, nothing in our intuition allows us to unite the two things. We cannot imagine a physical thing like a particle that has no absolute location like a wave. And we cannot comprehend what a wave that is propagated by electromagnetic fields would look like as a physical speck floating through space. The concept defies our attempts to visualize the properties of the two expressions of the same phenomenon together.

However, the unparalleled predictive abilities of the mathematical equations developed over the last century do not allow us to rid ourselves of the duality. Somehow the nature of light embodies both particle and wave, but neither precisely, at least as we have always understood particles and waves. The scientific community had to accept this fact, and many scientists resisted progress at first, but eventually no one could deny the raw data collected by the physicist. Light was like nothing we had ever seen in the natural sciences.

And still, through the work of de Broglie, the scientific community had to accept one more difficulty in this mysterious area of study. When we finally came to accept our understanding of the wave-particle duality of light, we found that all subatomic particles do the same thing. Not only photons behave this way; so do electrons and protons and neutrons and gluons and all the other -ons. Over the past few decades, scientists have consistently demonstrated wave-particle duality in larger and larger particles. Our understanding of all physicality, that which we experience daily and thought we knew all about long ago, has been shaken to the core.

In understanding Bohr's atom, we realized that all matter, in spite of its solidity, contains much more empty space than physical stuff. Now we find that the physical stuff that makes up such a small portion of the universe is not necessarily even physical stuff. It is just conglomerations and waves and bonds of energy. How did this happen?

According to Newton's second law of motion, "whenever we have a flow of energy, it's reasonable to expect that there will be an associated momentum – the two are the related time and space aspects of motion" (Hecht 45). With the development of the wave-theory of light in the beginning of the nineteenth century, Newton's concepts seemed to work well for only the practical, macroscopic world of our daily experience. The principles that determined the motion of baseballs flying through the air and blocks sliding down an inclined plane simply would not work at the microscopic level. In his work on the photoelectric effect, however, Einstein pulled Newton up from the ashes, and de Broglie's work put a tight mathematical explanation behind it. The frequency of a wave and the momentum of a particle are related to each other through Planck's constant, h. This infinitesimally small number made it possible to relate particles and waves.

The development of the Schrödinger-de Broglie wave theory did throw some light on the significance of Planck's celebrated constant h. Clearly, on the submicroscopic level of the atom, the concepts of "wave" and "particle," carried over from the macroscopic world of our daily experience, begin to break down. At the same time, it becomes as meaningful to speak of *wavelength* and *frequency* as somehow related to *momentum* and *energy*. The constant *h* merely takes care of the quantitative aspects of going over from the wave language to the particle language. In this sense, it occupies a role similar to that of the velocity of light in the formula $E=mc^2$: it is no more than a conversion factor. (March 216)

As referenced above, we find similar things happening at the massive scales of Einsteinian relativity as we do at the miniscule levels of photons. Einstein found that mass and energy as well as space and time were simply manifestations of the same phenomena, reconceptualized as mass-energy equivalence and spacetime. Only at speeds approaching the universe's built-in limit, *c*, the speed of light, do we find two things that were always assumed to be two things are actually two expressions of the same thing. Because *c* is so large, for much of human history an appreciable speed of *c* was unattainable. But since his time, Einstein's theoretical work has been confirmed by various experiments in which these speeds are achievable, like in the Large Hadron Collider particle accelerator.

In the same way, the smallness of *h* had left scientists completely unaware of the fact that light's wave and particle natures are really two expressions of the same phenomenon. "Just as the large value of *c*, the speed of light, obscures much of the true nature of space and time, the smallness of *h* obscures the wave-like aspects of matter in the day-to-day world" (Greene 105). In understanding this scientific fact, we are able to appreciate that electrons and protons are not just physical things carrying a charge, but they are energy itself: "the charges that constitute bulk matter" (Hecht 45). We have a world that somehow becomes a little less tangible as we go significantly down in size.

Certain philosophers would not be bothered by these developments. George Berkely, who died in 1753, maintained that matter did not exist, placing supreme importance on human perception. So long as a thing was not perceived, it did not properly exist, according to Berkely. Only in interaction with other life in the universe did a thing draw its existence. This type of thinking is what allows us to ask whether a tree that falls in the woods makes a sound or not. The logical difficulties of this philosophy were overcome by Berkely in that he believed that since there was an ultimate divinity that perceived all in the entire universe simultaneously, we would not expect our world to experience disappearance and reappearance based on whether or not someone was looking (Russell 647). God's observation grounded everything.

This is an interesting thought mused upon by the main character of Christopher Nolan's film, *Memento*. But after a brief philosophical daydream, Nolan's character states easily that the world goes on even while we are not looking. Modern man tends to agree with this practical conclusion, though we sometimes let our imaginations get the best of us.

As we see with Schrödinger's Cat, however, reality might not be so simple. Somehow, perception does play a key role in the minutest and briefest interactions in our world, an unimportant fact in our daily world, but one of supreme significance 'down below'. Perception determines and complicates how light behaves. Only as we generalize microscopic interactions can we start to focus on one particular aspect of light, ignoring the complications of the other contradictory aspects. One example of these generalizations is how we explain the practical optical properties of photons:

> When large numbers of particles are involved, probabilities approach certainties... In this way, the interference and diffraction patterns previously explained by waves can be interpreted as manifestations of particles. (Pedrotti and Pedrotti 5-6)

This is a fine perspective for the student of optics. When the mathematician can discard outliers through the overwhelming consequences of generalizations, statistical studies can continue, but the problem is that those outliers still exist. At the quantum level, all individual photons are outliers, completely incomprehensible in themselves, but the results of all of these infinitesimal interactions generalize into usable results and mathematical expressions relating waves and particles. Generalizations will do nicely for the pragmatist, but not for he who is interested in the realities of nature.

> In the new setting of ideas the distinction [between particles and waves] has vanished, because it was discovered that all particles have also wave properties, and *vice versa*. (Hecht 33)
> - Erwin Schrödinger

> The period during which light was "sometimes a wave and sometimes a particle" was a period of crisis – a period when something was wrong – and it ended only with the development of wave mechanics and the realization that light was a self-consistent entity different from both waves and particles. (Kuhn 115)

Gradually it became clear...that photons and electrons were neither waves nor particles, but something more complex than either. In attempting to explain physical phenomena, it is natural that we appeal to well-known physical models like waves and particles. As it turns out, however, the full intelligibility of a photon or an electron is not exhausted by either model. In certain situations, wavelike attributes may predominate; in other situations, particlelike attributes stand out. We can appeal to no simpler physical model that is adequate to handle all cases. (Pedrotti and Pedrotti 4)

Physicists still understand that they live in a world where nothing quite explains what they are witnessing in their experiments. The reality of light is too complex for us to apply to it any one concept or model. It defies categorization. Light is light, and that is about it. It cannot properly be called anything else, though we categorize different wavelengths across the spectrum. And light has laid bare the enigmas of the quantum world, those enigmas shared with light that compose all of the reality of our universe. In spite of all we gain through science, we still do not understand the basis of our natural world.

Meanwhile, physicists in their happy ignorance carry on using and applying quantum mechanics to great effect... Philosophical attitudes play little part in these efforts, beyond the elementary thought that physics ought to partake of old-time realism. (Lindley 209)

Old-time realism: that thing that allows science to push forward without understanding what it has already discovered, so long as what has been discovered continues to yield useful experimental results. The scientist may feel uncomfortable with the odd arbitrariness of physical laws; however, the mathematics that explain these scientific concepts are so refined and perfectly predictive that the pragmatism of science could hardly care less. Progress can still be made without fully understanding the paradoxes. It is only the least consolable theoretical scientists, many of whom are now working in string theory, who continue to ask the questions that do not yet have answers.

The strangeness science has unveiled in its exploration of light is duality:

Physicists use the term duality to describe theoretical models that appear to be different but nevertheless can be shown to describe exactly the same physics... Nontrivial examples of duality are those in which distinct descriptions of the same physical situation *do* yield different and complementary physical insights and mathematical methods of analysis. (Greene 298)

Dualities like the wave-particle duality of light are rare in science. Far more of our scientific understanding is seemingly straightforward in comparison. However, they may not be so uncommon in philosophy. For example, we can consider: "Ethical theories may be divided into two classes, according as they regard virtue as an end or a means" (Russell 178); however, it is not ridiculous to think of virtue as the means to accomplish the end of virtue. Many dualities, like these examples, blend the line of certainty and force the serious thinker, be he philosophic, religious, or scientific, to accept the fact that what he does not understand is hidden between contradictory ideas that somehow must be treated together.

If we want to seek more dualities outside of science, we should look no farther than religion. Christianity, it seems, is brimming over with dualities and enigmas in its beliefs and the language it uses to express eternal realities. Religion, and Christianity specifically, realized long ago what science only began to appreciate in the last century: reality is not just difficult to understand. It is incomprehensibly complex.

DUALITIES AND TRINITIES

Quantum mechanics and Einsteinian relativity uncovered some of the deepest mysteries of science. No longer are scientists only vexed with what is unknown, but now they have to contend with known concepts that do not make much sense. The data is unarguably precise, but the principles drawn from the data are unarguably complicated. We do not know what to do with what we know.

One of the natural consequences of quantum physics is the concept of uncertainty. In 1927, Werner Heisenberg proposed what is now known as Heisenberg's uncertainty principle, in which he expostulated and mathematically confirmed that only certain data can be determined about quantum particles at any one time. Practically, the quantum physicist can know the location of a photon with some certainty, but the more he knows the particle's location, the less he can know about how it is moving. If the scientist can determine how the photon is moving, he must necessarily be less sure of its location. Both velocity and location can be generalized, but the more precisely one is measured, the less able the scientist is to determine the other. When we know more about the particle, we know less about the wave, and vice versa.

There are not only practical experimental limits to what the physicist can know about elementary particles; Heisenberg's uncertainty principle goes even further by saying that theory does little to uncomplicate the issue. In the real world, we know that both location and velocity cannot be determined simultaneously, but theoretically we find the same thing. The mathematical formalism used to study one or the other quality of the photon does not allow the mathematician to study the other with precision. As we become certain of the one value, we become uncertain of the other. Ultimate truth is unknowable at the quantum level.

Naturally, non-scientists have latched onto this idea. In literature, critics may enjoy talking about one far-fetched interpretation, but interpreting the text one way closes the door to some other interpretations. Perhaps by assuming the author is communicating a critique of capitalism, the literary critic restricts her ability to also read into the allegory that would speak to the author's childhood. The average literary critic is less troubled by these consequences than the scientist. Scientists were happy to develop the uncertainty principle to deal with the quantum issue, but when philosophers began to apply uncertainty broadly and specifically to science, scientists rebelled against the consequence of ultimate unknowing. This rebellion, natural for the scientist, is manifest in the attempts of many to discover what underlies uncertainty, which would hopefully illuminate principles that will give a systematic explanation for the weirdness of quantum physics. String theorists are trying this.

The problem is that there is not yet an explanation, and in spite of the optimism of string theorists, man's ability to experimentally confirm what string theory proposes seems extremely far off in the future and potentially impossible. Until string theory or something like it is formally confirmed, "the uncertainty principle makes scientific knowledge itself less daunting to the nonscientists and more like the slippery, elusive kind of knowing we daily grapple with" (Lindley 7). That is not to say the scientific concepts are easier to grasp or the mathematics are less complex, but the non-specialist now feels like he has authority to weigh in on topics far outside of his ability. This unfounded confidence can be understood thusly:

> The uncertainty principle has become a catchphrase for the general difficulty, not just in science, of establishing untainted knowledge... When literary theorists assert that a text offers a variety of meanings, according to the tastes and prejudices of different readers, Heisenberg lurks in the background. (Lindley 7)

Of course, this was not the motivation for the uncertainty principle. Heisenberg simply created a formal understanding of the fact that one or another thing can be known about quantum particles, but not both at any single instance. His execution was scientific, though the application of his principle can be taken far outside the confines of science.

Similarly to Heisenberg's uncertainty principle, Schrödinger's Cat has found its way into popular and unscientific circles. The non-scientist feels free to expound upon the implications of perception and observation, even when the concepts as they are studied are not quite as the scientist defines them. But we can hardly blame the layman when scientists themselves remain in contention on the physical consequences of observation:

> The relevant literature is famously contentious and obscure. I believe it will remain so until someone constructs, within the formalism of quantum mechanics, an "observer," that is, a model entity whose states correspond to a recognizable caricature of conscious awareness. (Rosenblum and Kuttner 10)
> -Frank Wilczek, Physics Nobel Laureate

What Wilczek is requesting in the above quote is a scientific definition of observation, which necessarily includes a definition of that which observes, the observer. In Schrödinger's Cat, the observation, or lack thereof, is what determines the superimposed and final state of the cat. The superimposed state, the realization of various simultaneous and contradictory states, is unsatisfactory for the typical scientist, though few would argue the validity of the thought experiment.

The end of one line of reasoning concerning superimposed states is the proposition of nigh-infinite parallel universes, one for each superimposed possibility, of which only one combination is expressed in our universe, but the remainder of which would be yet one more fork in the road of an endless branching off of different realities. If this branching actually happened, each superimposition would generate a new universe. And each possible state of each elementary particle in our entire universe interacting at each specific moment of its existence with other particles doing the same multiplies into an incomprehensibly massive, yet quantifiably finite, number of possible universes. The honest scientist and the high school graduate can both quickly appreciate the impossible complexity of such a reality of realities, as did Augustine when critiquing Epicurus nearly two thousand years ago.

> The Epicurean [atomist] fantasy of innumerable worlds... asserts that these worlds come into being and then disintegrate through the fortuitous movements of atoms. (Augustine, City of God 434)

The innumerable worlds of modern parallel universes, however, would not disintegrate like that of the Epicureans but would continue having effect on the nigh-infinity of other possible configurations of elementary particles in each.

Within this bottomless pit of realities, the modern secularist contends that our universe is just one fortuitous example of physical stuff abiding by some physical laws to create our universe's actual physical reality. One is tempted to employ Augustine's word choice 'fantasy' to describe the mental leaps required to find comfort in such a formulation of our universe. When the secularist mocks the religious for leaning heavily on the crutch of theology, the religious has every right to mock the secularist for hiding behind theoretical legitimacy of endless probability. The mathematics may be sound, but, as the secularist would be happy to remind us, an intelligent human has no right to lay hold of mere possibilities without reasonable experience to defend the acceptance of the probabilities, and man has no experience or measurement that allows for parallel universes or scientifically definable 'ultimate observers'.

The scientist struggles, but the philosopher and theologian do not find the same restrictions when it comes to an 'ultimate observer'. As we have seen from philosopher George Berkely, he conceptualized the ultimate observer who would establish the continuity of a world otherwise plagued by tenuous existence. Berkley's observer will not do in our world, however, because the observer in the real modern quantum world must maintain quantum continuity and a single steady path out of all superimpositions while not at all collapsing Schrödinger's wave-function and disrupting the wave-nature of the photon. This observer would necessarily be a great source of enigma himself. The Christian theologian has come prepared for this eventuality. Augustine addresses God in all his divine enigma:

> Who then are you, my God? What, I ask, but God who is Lord? For who is the Lord but the Lord, or who is God but our God? Most high, utterly good, utterly powerful, most omnipotent, most merciful and most just, deeply hidden yet most intimately present, perfection of both beauty and strength, stable and incomprehensible, immutable yet changing all things, never new, never old, making everything new and leading the proud to be old without their knowledge; always active, always in repose, gathering to yourself but not in need, supporting and filling and protecting, creating and nurturing and bringing to maturity, searching even though to you nothing is lacking: you love without burning, you are jealous in a way that is free of anxiety, you repent without pain of regret, you are wrathful and remain tranquil. You will change without any change in your design. You recover what you find, yet have never lost. Never in any need, you rejoice in your gains; you are never avaricious, yet you require interest. We pay you more than you require so as to make you our debtor, yet who has anything which does not belong to you? You pay off debts, though owing nothing to anyone; you cancel debts and incur no loss. But in these words what have I said, my God, my life, my holy sweetness? What has anyone achieved in words when he speaks about you? Yet woe to those who are silent about you because, though loquacious with verbosity, they have nothing to say. (Augustine, Confessions 4-5)

That about sums up the divine enigma of Christianity. If you missed the profundity of what you just read, go back and read more slowly. Augustine touches on so much of the incomprehensible character of the Christian God in this one quote. Of course, as he himself recognizes at the end of the quote, in being careful to encapsulate the entire character of God, the philosopher did not really say anything. The contradictions of each sentence nullify any assertion made. And still, there is no other way to go about describing this god. The biblical scriptures confirm as much. And Augustine follows this thought by encouraging his reader:

If anyone finds your simultaneity beyond his understanding, it is
not for me to explain it. Let him be content to say "What is this?"
So too let him rejoice and delight in finding you who are beyond
discovery rather than fail to find you by supposing you to be
discoverable. (Augustine, Confessions 8)

Augustine recognizes that it would be impossible to explain the qualities
of the ultimate enigma. In describing one thing, he precludes another. By
determining one characteristic, he cannot know with any certainty the state
of another. We can only know in our unknowing, in pure experience.
Uncertainty prevails just as it defines the smallest physical building blocks of
our universe. Yet, all hope is not lost. We still live in the universe. We still
see it and test it and know it. We have come so far in our complex theories,
yet we have known certain things intrinsically from the beginning. Our
knowledge is now more refined, but we have always had some knowledge.

For what can be known about God is plain to them, because God
has shown it to them. For his invisible attributes, namely, his
eternal power and divine nature, have been clearly perceived, ever
since the creation of the world, in the things that have been made.
(Rom 1:19-20)

In the relentless continuation of enigma, however, we find that though
the divine nature is clearly expressed in his creation, that very creation in its
rough physicality masks the divine light. "Darkened is this world. Few have
insight here" (Dhammapada 13:174). Some guess that the difficulty arises
from the disparity between the physical and the spiritual components of the
world. "[Biological life] has, to be sure, a certain shadowy or symbolic
resemblance to [spiritual life]: but only the sort of resemblance there is
between a photo and a place, or a statue and a man" (Lewis, Mere Christianity
87). The physicality of biological life, the conglomeration of the energetic
bonds that compose all physical existence, masks and casts shadows against
the pure lights of the spiritual world.

These shadows are a minor theme in the Bible, repeated a number of
times to describe earthly religion and devotion. In the Christian conception,
this world is passing and is only a brief expression of eternal realities, of both
God and his spiritual creation. Behind the world, in the spiritual realm, we
find reality. "These are a shadow of the things to come, but the substance
belongs to Christ" (Col. 2:17). Man usually considers the spiritual realm of

ghosts and phantoms to be less real and certainly of less substance than our world, but the Bible tells us that the spiritual realities of creation align more closely with the divine source of creation, God of course being the ultimate reality. From this understanding the writer of Hebrews is able to deemphasize the work of the Levitical priests in saying, "They serve a copy and shadow of the heavenly things" (Heb. 8:5). The Christian understanding of the Old Covenant (the promises made to Abraham, Isaac, and Jacob, the law given to Moses, and all reconfirmed through David) is that it was merely a representation of the greater realities of heaven. The Jewish temple in Jerusalem was just a shadowy representation of the heavenly throne room. The rites of the Levitical priests were just practice for the ultimate tasks of devotion and worship in eternity.

Hinduism, in the understanding of the Vedas, would propose the same thing. Earth is just a shadow of the divine: "He of whom all this world is but the copy, who shakes things moveless, He, O men, is Indra" (Rig 2:12.9). Simply, physical creation cannot fully reflect the divine character that created it. A creation cannot fully contain its creator. Augustine suggests that the kingdom of God in heaven differs from the kingdom of God on earth "as widely as the sky from the earth," (and goes on to conceptualize it, unsurprisingly, in terms of light), as widely as "...the light of the Maker of the sun and moon from the light of the sun and moon" (Augustine, City of God 206).

If the divine cannot be fully expressed in our world, then naturally there will be certain aspects of divinity, of reality, that will remain unknown to us. "The secret things belong to the LORD our God, but the things that are revealed belong to us and to our children forever" (Deut. 29:29). This Old Testament verse is reminding the religiously devout to hold onto the truths illuminated in nature, in science, and in their own personal experience. It is man's duty and his honor to hold onto and seek out these truths. YHWH wants him to do just that. "It is the glory of God to conceal things, but the glory of kings is to search things out" (Prov. 25:2). In the Judaic scriptures, we are charged by the god of the universe to explore his creation and seek out his truths. He did not ban us from knowledge, just the tree of the knowledge of good and evil. He did not ban us from insight, just the realization of the potential of our own wicked powers.

All this religious enigma is complicated and frustrating. With every assertion that Christianity makes, it seems to dig itself a deeper hole. The critic and the skeptic justifiably push against the logical difficulties of the

Bible. But the logical weaknesses of Christianity only appear to be weak. They are actually enigma. In encompassing a greater swath of various truths, though it may begin to seem self-contradictory, the theology developed is actually shedding logical impossibilities by infusing itself with enigmatic complexity.

> If Christianity were something we were making up, of course we could make it easier. But it is not. We cannot compete, in simplicity, with people who are inventing religions. How could we? We are dealing with Fact. Of course anyone can be simple if he has no facts to bother about. (Lewis, Mere Christianity 90)

If science wished to be simple, it would quickly rid itself of the facts of quantum physics and relativity; it would throw away the study of light before Newton could jot down the first word of *Opticks*. But that is not what man desires. We study nature to learn what reality is like. We do not ignore facts when they are seen as facts. Simplicity is not what we want. We want truth, regardless of the consequences, regardless of the enigma. Let us embrace the enigmas we find and all their consequences.

One consequence is for the religious. If a religious person tells an unbeliever that he will show him the face of God, he is either intentionally lying or ignorant of the truth. The religious, even the prophets, even Moses who communed with God more intimately than any mortal, have not perceived God as he is in his fullness. The Qur'an confirms this fact about our observation of the ultimate observer: "No vision can take Him in, but He takes in all vision. He is the All Subtle, the All Aware" (6:103). Likewise, the apostle Paul confirms Muhammad's statement in describing God as he "who dwells in unapproachable light, whom no one has ever seen or can see" (1 Tim. 6:16). C.S. Lewis asks in his masterful allegorical work *Till We Have Faces* how the divine would even be able to commune with man while man's vision is blurred by the constraints of physical and spiritual imperfection that detract from our reality: "How can [the gods] meet us face to face till we have faces?" (Lewis, Till We Have Faces 294). Still more, Thoreau continues on this line of argument, on our inability to perceive the reality we crave: "The light which puts out our eyes is darkness to us. Only that day dawns to which we are awake. There is more day to dawn. The sun is but a morning star" (Thoreau 351).

There is a light. This light is more real than any light we have ever seen. By this light will we see things we have never perceived. But until we attain the 'enlightenment' that would allow us to perceive higher things, until we attain the understanding to grasp the impossible enigmas of our reality, that light which will someday make everything clear and bright can only blind us in our current state of ignorance.

> Yet among the mature we do impart wisdom, although it is not a wisdom of this age or of the rulers of this age, who are doomed to pass away. But we impart a secret and hidden wisdom of God, which God decreed before the ages for our glory... These things God has revealed to us through the Spirit. For the Spirit searches everything, even the depths of God. (1 Cor. 2:6-7, 10)

If through the above verses a religious person is tempted to claim supernatural insight into the nature of God and his reality and his creation, he would do well to consider the following:

> If there is a prophet among you, I the LORD make myself known to him in a vision; I speak with him in a dream. Not so with my servant Moses. He is faithful in all my house. With him I speak mouth to mouth, clearly, and not in riddles, and he beholds the form of the LORD. (Num. 12:6-8)

If anyone had a claim to understand the divine realities of our world, according to the Judaic scriptures, that man was Moses. He had ascended the mountain and entered the tabernacle. He spoke with YHWH and saw his form passing. He was God's mouthpiece for the people, and he pleaded with God to bear with the Israelites, time and time again. Moses *knew* YHWH, he knew the divine, more than any mortal ever. He spoke to YHWH face to face, as friends do. Yet, at 120 years old, after following YHWH devoutly for 40 years, Moses says:

> O Lord GOD, you have only begun to show your servant your greatness and your mighty hand. For what god is there in heaven or on earth who can do such works and mighty acts as yours? (Deut. 3:24)

146

Moses, the one man who had more of a right to claim that he knew the character of the divine, instead claims that he had only seen the very tip of the iceberg.

It could take an entire book in itself to treat the Christian concept of the trinity, which is simply not the intention of this current work. But anyone discussing the incomprehensible enigmas of the Christian God would be remiss in not at least introducing the topic. So please be understanding of the brevity of this discussion. Those who are interested in the concept are encouraged to research it on their own.

The trinity is a Christian understanding of the triune God; the Father, the Son, and the Holy Spirit: three manifestations of one god. These are somehow three independent beings unified in perfect communion, blurring the lines and distinctions that separate their individual expressions in our world. The existence and unity of these three beings must not betray the perfection necessary for eternity, as eternity cannot abide change or imperfection lest time eek into the divine infinities.

The Father abides over all existence and created the physical world through his Word, which we are told was his Son, the primary expression of the Father's divinity in the physical world, creating it and coming in human form to redeem it. On earth, Jesus was fully man and fully God, something beyond our understanding in itself. The Spirit maintains a continual connection between the spiritual realm of the Father in heaven and the physical realm of the Word on earth. The Spirit was there in the beginning hovering over the void and formless creation and comes upon the disciples when Jesus is resurrected and returned to the spiritual reality of heaven.

The existence of the trinity is one of the primary enigmas of the Christian faith, and it is also a barrier to many who would otherwise be tempted to accept the teachings of the Bible. Jewish and Islamic scholars viciously oppose the heresy of maintaining a belief that God is anything but the supreme, *one* god, YHWH or Allah.

What the Christian believes, however, is just that. Of course, she also believes in the three persons of God. She believes in paradox. The same flexibility involved with accepting the wave-particle duality of light will allow the intellectual to accept the enigma of the trinity. Although the evidence for each is very different from the other, the logic is not.

We have a being which expresses itself differently in different situations. Physically, we can understand this single phenomenon in multiple ways; however, no person with a mature understanding of the concept will actually think that the phenomenon is one thing at one time and another at another. Instead, as light is both wave and particle but neither precisely, so God is Father, Son, and Spirit, though not a single one of those three encapsulate the unity and reality of God. Depending on how we are 'measuring' the divine, it will express different models, just like the light that pervades and composes the universe.

The trinity, as the wave-particle duality of light, is a helpful construct for the earth-bound and time-constrained human to understand the practicalities of an incomprehensible reality, and only in an enlightened, redeemed state do we have hope of witnessing the unity of the divine light in its totality.

The trinity is difficult, for similar reasons why quantum physics and relativity are difficult, and yet the faith claimed by Christians is incredibly simple. In considering the scriptures of this enigmatic faith, one can say:

> I therefore decided to give attention to the holy scriptures and to find out what they were like. And this is what met me: something neither open to the proud nor laid bare to mere children; a text lowly to the beginner but, on further reading, of mountainous difficulty and enveloped in mysteries. (Augustine, Confessions 40)

Even a child can read and benefit from the Bible, but if we were hoping the scriptures would bring clarity, we must at this point be sadly disappointed. It seems to be the very source of great difficulty, and the text itself is enigma to its core, in its very wording.

Italo Calvino, a prolific novelist in his time and a man of great descriptive ability, penned this quote: "But I do not believe totality can be contained in language; my problem is what remains outside, the unwritten, the unwritable" (Calvino 181). Language seems, even to great writers, to lack the qualities needed to encapsulate all that they would like to capture.

> There is always something deeper than anything said – something of which all human, all divine words, figures, pictures, motion-forms, are but the outer laminar spheres through which the central reality shines more or less plainly. Light itself is but the poor outside form of a deeper, better thing, namely, life. (MacDonald 209)

As we find the written and spoken word to be of little help in understanding the intentions of the speaker, let alone the enigmas of an impossibly complex reality, it is no wonder that we are generally dissatisfied with man's state. We have this innate desire to seek and to know, and yet there is a significant portion of our existence that is shrouded behind our own physicality; we are unable to perceive the divine light that ought to be flowing freely through us as divine energies. The main character and narrator of Lewis' *Till We Have Faces* approaches the end of her text with this:

> I ended my first book with the words *no answer.* I know now, Lord, why you utter no answer. You are yourself the answer. Before your face questions die away. What other answer would suffice? Only words, words; to be led out to battle against other words. (Lewis, Till We Have Faces 308)

For the one who finds himself wrapped up in the frustrations of trying to conceptualize the enigmas of light and the divine, his peace and answer is only light itself. There can be no other answer. Nothing encapsulates its totality, its reality. Wave is only part. Particle is only part. Electromagnetic disturbance is only part. Human wisdom is only part. Human will is only part. Divine presence is only part. Father, Son, and Spirit are only parts.

The only answer is *light* itself.

We hold out hope that we will one day see through the enigma and understand light as it is, physically, philosophically, and theologically. We wait until we are able to perceive it, to digest it fully. "The natural man is as an infant in Christ and a drinker of milk, until, he is strengthened for solid food, and acquires eyesight strong enough to face the sun" (Augustine, Confessions 286). Just so:

> For we know in part and we prophesy in part, but when the perfect comes, the partial will pass away. When I was a child, I spoke like a child, I thought like a child, I reasoned like a child. When I became a man, I gave up childish ways. For now we see in a mirror dimly, but then face to face. Now I know in part; then I shall know fully, even as I have been fully known. (1 Cor 13:9-12)

We may yet see things clearly, but the first steps we take toward perceiving the light are not as inspiring as we might hope. However, through those elementary steps, through that elementary understanding, we find ourselves dropped into the middle of a divine experience that eclipses any pure knowledge we could hope to attain.

> We know life only as light; it is the life in us that makes us see. All the growth of the Christian is the more and more life he is receiving. At first his religion may hardly be distinguishable from the mere prudent desire to save his soul; but at last he loses that very soul in the glory of love, and so saves it; self becomes but the cloud on which the white light of God divides into harmonies unspeakable. (MacDonald 151)

The Christian does far more than believe he should go to church and not swear too much and be nice to his grandma. The Christian believes and actively works toward forming himself into the person through which God's divine light can shine all the greater. Only in becoming the instrument through which divine light and perfect will enter into our dark world will the mortal finally perceive the fullness of that which he was seeking. Only in living into the magnitude of his glorified and eternal state will man attain that fullness. Eternity is in every passing moment; light gives access to that eternal present.

RELATIVE & ABSOLUTE

The American road trip: A long, straight road in Montana, going on and on and on for what seems like hours without much change. A range of mountains in the rearview shrinks, but so slowly that it disappears without you noticing minute by minute. There is another range of mountains ahead that you just cannot gain on, set at a far distance that is maintained for an hour or more. Are you making any progress? Nothing moves, though you are constantly traveling at 80 miles per hour.

Regardless of what your perspective is on how much ground you are covering in reference to the mountains in the distance, from minute to minute, as the time passes, you continue to travel at 80 miles per hour. Still, things in the distance, big and small, do not seem to move. This phenomenon is parallax.

Parallax is an optical concept that measures the angle of change between two points of reference. There will be greater parallax between objects two feet apart than objects two miles apart. This is because when you take two steps directly to the right, objects two feet away can move from your center of vision to the far left. Two steps to the right hardly changes your perspective of the object two miles away. Your angle of view has changed by a far smaller degree in the second case.

The mountains in the distance change so slowly that you cannot perceive the change from minute to minute. Contrarily, objects up close fly by. Telephone poles slowly approach the car, speeding up and whizzing by the car as you get very close. In the passing of these objects, you remind yourself that you are in fact driving quickly down the road, making the progress you expect. The whole time, everything that is stationary by the road or ahead in the distance are properly moving at 80 miles per hour relative to your car's motion. Telephone poles pass by at 80 miles per hour. A broken-down car off the berm passes by at 80 miles per hour. A farmhouse a mile off the highway passes by at 80 miles per hour. The mountains way off to your left pass by at 80 miles per hour, regardless of your perspective.

The speeding car coming up on your left, however, is not passing your car at 80 miles per hour. The Porsche seems to be going about 100 miles per hour. When it passes you, you see it going about 20 miles per hour faster than you. It takes the same time to pass you as a car going 20 miles per hour would when you are at a full stop.

Comparing the speeds of the cars requires the concept of relative velocity. The Porsche does not pass so quickly as the telephone poles flying by at 80 miles per hour. It has a slower speed relative to your car even though it is moving faster. Conversely, a car on the opposite side of the highway will have a much higher relative speed. Although telephone poles pass by at 80 miles per hour, the car travelling the opposite direction, also driving at 80 miles per hour, will seem to fly past you at 160 miles per hour, double the speed limit. Luckily for that driver, by the policeman with the radar gun on the side of the road the car will properly be clocked at 80 miles per hour instead of doubling the speed limit.

It is fairly easy for us to picture the concept of relative velocity on earth. Our daily experience informs what we see, and we are not at all surprised by the cars on the opposite side of the road travelling at outrageous speeds. Our perspective accounts for relative speeds naturally. We have the mental constructs to understand what we are seeing.

As one experiences in an elevator, constant and non-accelerating motion cannot be perceived if one has no special reference to ground her movement. As the elevator moves up floor to floor, we feel the downward force of the motion upward until the elevator reaches a constant speed. At that constant speed, without acceleration, we feel as if we were standing on level ground. We move in unison with the inside of the elevator, so we have no visual indications that we are moving at all, aside from the dinging of the floor number as we go up and up. We still feel gravity, the accelerative force of earth holding our feet onto the floor of the elevator, but we feel no other motion until the elevator slows as it approaches our floor. At that point, we feel an upward force, or a lightening, counteracting the force of gravity.

Likewise, in a car, train, or plane, after a constant speed is achieved, we do not feel the effects of acceleration. We can walk through a plane while it is travelling at 500 miles per hour as if we were walking on flat earth, all the while feeling the accelerative force of gravity pulling us down as we expect. These effects are the consequence of Galileo's conception of relativity, upon which Einstein developed his own theories.

Now, we can apply these concepts to a mental picture of space travel to the same effect. Suppose one astronaut, Jennifer, has been expelled from her spacecraft. She is in the utter blackness of space with no star, planet, comet, or any other thing within lightyears of her location. No perceptible parallax or objects whizzing by can help her determine her speed or direction of movement. Although she travels at a particular speed, she has no object to reference to determine that speed.

She is travelling at a constant velocity, experiencing no acceleration, no speeding up or slowing down or changing of direction. Even more, with no planet or star close enough to have an appreciable force on her through gravity, Jennifer experiences no force on her that she can perceive. With no acceleration, no special references, no gravity to act upon her motion, she has no means to determine her movement. For all intents and purposes, Jennifer might as well be stationary because her constant velocity is having no effect on her perception. In fact, she guesses that she is completely still, though she is actually travelling a quarter the speed of light away from earth.

In an odd coincidence, another space traveler named Kif has also been expelled from his spacecraft travelling with the exact opposite velocity, that is, in the opposite direction at the same speed as Jennifer. Kif is experiencing the same utter motionlessness that Jennifer is, although he is also travelling at the constant speed of a quarter of the speed of light toward earth.

Neil is yet one more astronaut, who was marooned in space by his crew when he tried to stage a mutiny. Neil is 'motionless' at the exact middle of these two space cadets, and he sees them both as they come and go, travelling at one quarter the speed of light.

From Neil's perspective, the motion of Kif and Jennifer are exactly the same, travelling in two different directions. However, Jennifer does not witness the same thing. Experiencing no acceleration on her person and having no other objects to reference for her own motion, Jennifer assumed herself to be motionless. What she sees as these three pass is this: Kif was travelling toward her at half the speed of light and Neil was dallying in her direction at only a quarter of the speed of light. As they pass, Kif speeds away, and Neil disappears into the distance in twice the time it takes Kif to go out of sight. Kif sees the same thing in reverse; Jennifer speeding past at half the speed of light and Neil going half her speed.

What the above thought experiment does, however, is to assume that we have an actual, stationary observer in Neil. In reality, each castaway has just as much claim that he or she is at rest while the other two are travelling at extraordinary speeds. Without any other thing to standardize their motion, they can all claim to experience the complete non-accelerative state of rest. In saying that Neil was stationary, we assumed that he was stationary in relation to earth; however, we cannot refer to space absolutely by saying that Neil remained stationary compared to something else, except by making that reference explicit. Space is only a measurement of the distance between things. It is not absolute in itself. We cannot speak as if it is, if we intend to remain scientifically grounded. And as we will soon see, space is not as generally absolute as humanity supposed up through the end of the nineteenth century.

We see in the above thought experiment that no one has a 'privileged' view of space and motion. Einstein reemphasizes Galileo's relativity in saying that no observer could lay claim to being at rest while another was moving, or vice versa. Instead, all observers can only speak about their motion relative to that of other objects. All that Jennifer could say was that Kif and she were coming together and separating at half the speed of light, not that she was still and he was moving or the other way around or anything in the middle where they each traveled at speeds complentarily equaling half the speed of light.

Like Galileo's theory, Einstein's theory of special relativity tells us that no observer has a privileged perspective, only in Einstein's theory, relativity refers to the speed of light. Special relativity reveals that "any and all observers will agree that light travels at 670 million miles per hour *regardless of benchmarks for comparison*" (Greene 31); that is, no matter what speed an observer is travelling, she will always perceive light to be approaching and departing her at the speed of light, as if she was not moving at all. At first blush, to those unfamiliar with relativity, this may seem like a fine proposition, but the logical and mathematical consequences of such a concept become very troubling and reveal one more enigma tied to light.

Let us return to our space travel thought experiment but alter it slightly. In this new example, Jennifer and Kif had been travelling in the same spacecraft; Jennifer was expelled from the ship at half of the speed of light, and Kif, for some reason, was expelled from the same ship in the same direction at the same speed one minute later. In between the two, 30 seconds after Jennifer and 30 seconds before Kif, a pulsing light was released from the ship with the same velocity.

Now remember, special relativity, which has been confirmed mathematically and experimentally with unparalleled accuracy and which we have no reason to doubt (besides its weird consequences), tells us that light travels at the same speed for all observers regardless of their speeds relative to other benchmarks. In this example, the speed of light is the same for

Jennifer and Kif, as it is for Neil. They all perceive light travelling at 299,792,458 meters per second relative to their speed, as if each were still. More, each has to recognize the legitimacy of the others' perspectives, though they may have trouble accepting the consequences.

This idea is counterintuitive to the relative motion we considered on the 'American Road Trip'. Typically, things travelling in the same direction will experience a lesser relative speed, and things travelling in opposite directions will experience higher relative speeds. However, light does not abide by this law of relative motion. Light travels at a constant speed for all observers at all times, but the result of this feature of light is that it thus travels at two different speeds when the astronauts compare their experiences with each other. The astronauts know, however, that light does not travel at two different speeds according to relativity, so instead, the components of speed, *time* and *distance*, are what flex to unify their experiences.

If Neil could somehow perceive the light that Jennifer and Kif see, he would see light travelling at 449,688,687 meters per second (1.5 times the speed of light) as the pulses approach him and 149,896,299 meters per second (half the speed of light) as it passes and sails into the distance. Neil would add the speed of the pulsing device to the speed of the light it emits as it approached, and he would subtract when it passed. But Neil does not do this because of Einstein's theory of special relativity. To Neil, the light emitted as the pulses approach travels at 299,792,458 meters per second toward him, and after the light passes him and recedes into the blackness of space, the light emitted in his direction still travels at 299,792,458 meters per second toward him, regardless of the relative speed of the object emitting that light.

Whether or not it is already clear to the reader, the light that pulsed in between Jennifer and Kif illuminates different consequences for those two than for Neil. Jennifer and Kif, at rest relative to each other and the light equidistant between them, since they travel through space with the same velocity, will agree that they perceive the same flash of the light at the same time. The light flashed when they were equidistant from it and light traverses those distances with the same speed relative to their speed, so the time it takes to reach each will likewise be the same.

Neil sees something else. The light flashes when it is directly beside Neil and travels toward Jennifer and Kif at 299,792,458 meters per second relative to Neil; however, Kif is speeding toward the pulse of light at half that speed and Jennifer is speeding away at the same speed. Thus, from Neil's perspective, the light reaches Kif before it reaches Jennifer. Neil does not and should not agree with the simultaneity that Jennifer and Kif experience. Light, in behaving the same for him (i.e., travelling the same speed relative to his velocity), interacts with objects in relative motion to him differently than those objects perceive it interacting with themselves. All these observations are valid and true and scientific, though seemingly contradictory. We account for the differences in simultaneity by altering the distance and time it took the light to travel in the different observations so that its speed can remain constant.

In spite of the differences of perception, special relativity provides a mathematical understanding that formalizes the validity of the enigmas centered around the speed of light. "In relativity, although observers may disagree, events retain a distinct and unarguable physicality... Special relativity accounts for the discrepancies... An underlying objectivity persists" (Lindley 132). When Einstein proposed special relativity, the concepts that the theory's mathematics illuminated seemed non-sensical. Truth be told, they still do.

Naturally with the discrepancy of simultaneity, special relativity also shifts the perception of distance and time when objects approach the speed of light. If a spaceship passed earth near the speed of light, the craft would appear linearly squished to us earthlings though it would appear normal for observers inside the ship. So too would earth appear shorter to the astronauts aboard the spaceship and normal to us. The relative speed between earth and the ship affect observers on both the same way, and both observations are valid.

Space itself constricts into an abbreviated expression when we perceive physical objects travelling near the speed of light. Likewise, time passes more slowly for objects travelling near the speed of light relative to a 'grounded' observer. Distance and time, the components of the measurement of velocity ($v=d/t$), are altered when speeds approach the speed of light, for the speed of light does not change. Einstein understood this to mean that space and time were not different dimensions in our universe as we always assumed, but actually two components of the same dimension of *spacetime*.

Spacetime is the means through which the differing observations of Jennifer, Kif, and Neil are justified to be correct for each observer.

Relativity, to be sure, allowed for differing perspectives, but the whole point of [Einstein's] theory was that it allowed apparently contradictory observations to be reconciled in a way that all observers could accept. (Lindley 5-6)

Through special relativity we can accept what Neil sees and accept what Jennifer and Kif see, even though their measurements and experiences do not agree. We have well-reasoned mathematical arguments that express exactly why the measurements are different, which allow for a certain, old-fashioned objectivity to persist. Reformulating our concepts of space and time into one spacetime allows us to understand the differing perspectives. Einstein saved and revigorated Newtonian physics by explaining why Newtonian physics failed at great speeds and under great forces. Einstein's relativity kept formal laws in place where objectivity seemed to crumble.

At the beginning of the twentieth century, relativity simultaneously created and solved the enigmas of objects travelling near the speed of light. Einstein's theories were revolutionary and have since been confirmed with great accuracy. Relativity upends our understanding of space, time, mass, and energy, and it all centers around the speed of light as a conversion factor and an intrinsic 'speed limit' of the universe. The speed of light became supremely important in Einstein's theories, but the problems solved by our appreciation for the speed of light have only illuminated more issues. Chief among these issues is how we are supposed to unify Einstein's relativity with the quantum understanding of atomic particles. These sets of theories defy unification.

Well formulated physical problems elicit nonsensical answers when the equations of both these theories are commingled. The nonsense often takes the form of a prediction that the quantum-mechanical probability for some process is not 20 percent or 73 percent or 91 percent but *infinity*. What in the world does a probability greater than one mean, let alone one that is infinite? (Greene 118)

String theory attempts to reconcile the difficulties of unifying quantum and relativity theories by setting a lower limit to the universe, using Planck's constant as the smallest factor, just as the speed of light is used as the largest factor. By establishing a lower limit, we rid ourselves of the mathematical difficulties of physical infinities and eternities. But light itself remains the limit, something unattainable for all other physicality in our universe. All physical matter can approach the speed of light, even though weird stuff starts to happen the closer to the speed of light we approach. But we still cannot get physical matter to travel at the speed of light. It is the unattainable limit of all physical matter; that is, of course, besides photons. A photon, the smallest bit of light and matter possible, behaves like nothing else at great speeds or down on the quantum level.

> A crucial difference between particles like electrons and neutrons and particles like photons is that the latter has zero rest mass... Thus, while nonzero rest-mass particles like electrons have a limiting speed of c... zero rest-mass particles like photons must travel with the constant speed c. (Pedrotti and Pedrotti 4)

Light must travel at the speed of light or it is not light. A photon must travel at the speed of light (slightly lower than 299,792,458 meters per second in a medium like water or glass and perhaps at the edge of the universe due to gravitational retardation) or else the photon does not exist. So long as no particle intervenes in a photons path, light continues to travel at the speed of light, ad infinitum. Its speed is intrinsically tied to its mass and its very existence. There is no point in trying to get a hold of a photon and putting it on a scale to measure its weight. If we could stop it on a scale, it would cease to exist and the scale would register no mass.

As established by special relativity, when objects approach the speed of light, space and time constrict into an abbreviated expression compared to the balance of the physical world. If we could launch astronauts away from earth near the speed of light, upon their return they would have experienced less time passing than those who stayed on earth. If a twin was launched into space like this, upon his return, he would be younger than his brother. This is difficult to accept intellectually, but the phenomenon has been confirmed by the extended life of highly unstable atoms in particle accelerators. Atoms which usually break apart in a fraction of a second can stay together much longer if travelling near the speed of light. This is because we see time passing more slowly for the atom travelling at great speeds relative to the physicist observing it. The atom breaks apart in the same amount of time as it always does, it is just that time passes more slowly near the speed of light.

One way to understand this feature of special relativity is to consider the ability and necessity of an object traveling through spacetime, space and time at once. Near the universe's speed limit, travelling through *space* with great rapidity, physical matter has less ability to travel through *time*. All its 'speed' is used up on space, leaving none for time. Conversely, if an object is at rest relative to all other matter surrounding it, it does not travel through space at all, leaving all of its movement through spacetime to be consumed by time. We do not usually perceive the expression of this duality because of the hugeness of the speed of light. Relative to the speed of light, everything normally travels very, very slowly through space relative to us, so all matter typically travels primarily through time, all at essentially the same speed, time passing for all of us at roughly the same rate. However,

> Something travelling at light speed through space will have no speed left for motion through time. Thus light does not get old; a photon that emerged from the big bang is the same age today as it was then. There is no passage of time at light speed. (Greene 51)

The space-travelling twin launched from earth near light speed and eventually returning will have experienced less time than his earthbound sibling. The astronaut will be younger. If he could somehow travel at the speed of light, he would undergo no passage of time at all. Upon his return to earth, it would be to him as if he never left. He would be the exact same age he was when he left compared to his twin who aged on earth in the interim. At light speed, an inaccessible hypothetical, no time passes relative to the balance of the universe. Light speed, for all intents and purposes, is the gateway to eternity.

This ends the treatment of the physics of special relativity that will be given in the current text. The topic has hardly been broached, and only a superficial explanation has been provided for a topic that has incredible depth. Hopefully, the consequences of the theory have been expounded sufficiently to establish a connection with the philosophical and theological correlations to follow; however, the unsatisfied or curious reader is encouraged to pursue his or her own research on this topic. The resources quoted in this chapter will serve as only a starting point from which extremely complex iterations of the theory might be presented for the serious skeptic and student of science and mathematics. Only when the consequences of relativity are completely believed will the correlations with philosophy and theology be fully appreciated. These correlations shed light on the very fabric of the reality of our beautifully complex universe.

ETERNITY AND INFINITY

Time is a curious thing. All humanity takes the passage of time for granted. Our daily experience makes what we perceive as time extremely clear and all but inescapable. We can hardly conceive of anything outside of the constraints of time. All that we know came to us in time, and all that we have forgotten has passed with time. Time pervades every ounce of our experience.

Not until we could send particles flying through an accelerator could we truly appreciate Einstein's theory of relativity and the ability to contract and slow time by means of ridiculous speeds. Not until humanity was armed with modern physics has man been able to understand that time itself and its passage through our universe is controllable and limited in scope.

But in finding the limits of time, man also found the infinities he dreamed about since he first conceived of the divine. In the speed of light, in the natural speed limit of our universe, man gains a glimpse into eternity. At the speed of light, the only speed it will ever travel, a photon acquires a certain physicality and timelessness, which ceases the moment it encounters and is absorbed by other matter, the moment it ceases to be light. And in approaching the speed of light, by testing the intrinsic speed limit of our universe, other matter gains access to a timelessness otherwise inaccessible to our physical world.

To the theologian, and even to the more imaginative philosopher, this access into timelessness ought to come as no surprise. Artists have imagined for centuries what ability they might have to access eternity. In their words, poets have sought to immortalize their beloved. Always in our understanding of the divine, eternity is never more than a step behind. Infinity calls to us. It is a concept with no actuality in our personal experience, yet, in a way, we all experience it in our expectations of the continuity of divine love and justice or the nature of the human soul or the perpetual laws of physics which bind our universe as one conceivable whole through all time. We know time goes on apart from us. We know, therefore, that the totality of time is beyond us, beyond our comprehension. Eternity is our explanation. But whatever form that explanation takes, certain attributes of time are ubiquitous:

> Things which happen under the condition of time are in the future,
> not yet in being, or in the present, already existing, or in the past,

no longer in being. But God comprehends all these in a stable and eternal present. (Augustine, City of God 452)

The theologian, truly anyone who imagines some eternity, even the scientist who conceives of some sort of consciousness travelling at the speed of light, must describe time in the above terms. Time divides our universe into threes. There is the past, that which is dead under the weight of ever successive passing moments. There is the future, that which encapsulates the imaginative possibilities and contingencies of all that we have experienced and are experiencing. And there is the present, that which *is* already, all that ever *is*, all that we can truly say exists. The present is an infinitesimally thin portion of time at which everything acquires its reality. Only in the present is the eternal accessible.

> Then it may compare eternity with temporal successiveness which never has any constancy, and will see there is no comparison possible. It will see that a long time is long only because constituted of many successive movements which cannot be simultaneously extended. In the eternal, nothing is transient, but the whole is present. (Augustine, Confessions 228)

Here is the crux of eternity, the importance of which simply cannot be overlooked. The child, in his conception of eternity, sees all of his own successive moments, those that were and those to come. He adds this totality with that of his parents. If possible, he may even imagine the lives of his grandparents and great-grandparents. He may be able to see his children in the future and the lives of their children. He may understand the gravity of history and be able to conceptualize Franklin D. Roosevelt and Napoleon and Charlemagne and Caesar. He might even imagine what God was up to at the beginning of creation. But the child, in his childish understanding of time, will necessarily imagine eternity as the compilation of all of those successive moments. The child understands eternity as *all* of history. God himself is confined within the history of reality and the passage of time from this perspective.

Serious science, philosophy, and theology teach something different. The mature thinker does not conceive of eternity as being composed of infinite successive moments. Instead, she understands that eternity can only be eternity if it does not undergo the inescapable ravages of time. Eternity is not defined by time, but it is defined by time's absence. Eternity is not measured by unlimited change, but by changelessness. Eternity is not *all time;* it is the opposite of time.

Humankind has no other category for eternity except that which it is. There is no comparison. We cannot truly understand it allegorically or metaphorically. Nothing from our everyday, mundane, physical experience tells us what infinity might be like. Only in our imaginations, in the actualization of invisible realities, do we see what eternity is: "For the things that are seen are transient, but the things that are unseen are eternal" (2 Cor 4:18). No truly scientific observation is eternal. Nothing physical abides. It all fades with time, literally or in our understanding. But that which lacks time-conditioned physicality lives on. Therein lies the hope of the poets who sought to immortalize their ladies in meter and rhyme. Therein lies the whole of our understanding of the divine:

> But God has no history. He is too completely and utterly real to have one. For, of course, to have a history means losing part of your reality (because it has already slipped away into the past) and not yet having another part (because it is still in the future): in fact having nothing but the tiny little present, which has gone before you can speak about it. (Lewis, Mere Christianity 92)

There is not an Old Testament and a New Testament God, one of wrath and one of love, as so many like to say. Insofar as any god is actually divine, that god's divine characteristics ought to remain the same eternally. There can be no change in a being like God. Divinity is eternal, and thusly, eternally present and abiding and unchanging.

It should come as no surprise that man for millennia has assumed access into this eternity, the infinities of time and divine character, would come through light. Light embodies that timelessness. In this spirit, we see statements such as: "Gods, we have gone to light... We have become immortal" (White Yajur 18:29). Likewise, the divinities of many religions describe themselves and all life in terms of immortality:

> Never have I not existed,
> nor you, nor these kings;
> and never in the future
> shall we cease to exist. (Gita 2:12)

Even in popular culture, the religions of the east influence our concept of time and eternity. In the internationally successful movie, *Crouching Tiger Hidden Dragon*, one of the main characters, Li Mu Bai, describes his meditation and a glimpse he had into the light of the infinite:

During my mediation training... I came to a place of deep silence...
I was surrounded by light... Time and space disappeared. I had
come to a place my master had never told me about.

The divine light does not experience the limits of time because it is apart from time; it is infinite in time and space for it created all time and space. To describe divinity, to describe eternity, we cannot use terms particular to time, though it seems man cannot help but to use terms particular to light: "There is no time for the Light of Lights prior to everything other than the Light of Lights, for time itself is also one of the things other than the Light of Lights" (Suhrawardi 115).

These concepts are supremely well-developed in the Christian tradition, both in the text of the Bible and in the interpretations of later theologians. In the prophetic writings of Revelation, the writer introduces the supreme divinity stating, "I am the Alpha and Omega... who is and who was and who is to come, the Almighty" (Rev. 1:8). This divinity encapsulates totality; he predates and transcends all and will abide beyond the extinction of everything. He is immortal. He is infinite. He is eternal. He is the beginning and end. And thus, necessarily, he is ever-present in time and space. He told as much to Moses in the encounter at the burning bush:

Then Moses said to God, "If I come to the people of Israel and say
to them, 'The God of your fathers has sent me to you,' and they ask
me, 'What is his name?' what shall I say to them?" God said to
Moses, "I AM WHO I AM." And he said, "Say this to the people of
Israel: 'I AM has sent me to you.'" (Ex. 3:13-14)

The text of the Pentateuch does not make it clear why the people of Israel would have recognized this name, I AM, or YHWH. This was either already a common way to refer to the god of the Israelites or the people were expected to recognize the legitimacy of Moses' claim based on the logical consequences of such a name. Almost certainly, the name was already in use, so we have to be careful to understand the verse below:

God spoke to Moses and said to him, "I am the LORD. I appeared to Abraham, to Isaac, and to Jacob, as God Almighty, but by my name [YHWH] I did not make myself known to them." (Ex. 6:2)

It would seem that in his discussion with God at the burning bush, Moses was not necessarily the first to hear his name, YHWH, but was simply the first to understand the profundity of this name. The text allows for either reading, though history may not, but regardless of which may be true, the modern reader should pause to try to appreciate the depth of the meaning of 'YHWH' as Moses presumably did.

In the New Testament, Jesus is recorded as saying, "Truly, truly, I say to you, before Abraham was, I am" (John 8:58). Jesus was talking about himself but referred to himself just as YHWH had referred to his divine self. In this statement, Jesus is either making the boldest claim ever stated by man, that he was one with the supreme divinity of the universe, or he was uttering the highest blasphemy. It should be no wonder that the religious leaders in Jerusalem wanted Jesus dead and gone. Jesus claimed to be YHWH himself. This kind of heresy can hardly be tolerated by any established religious order.

However, Jesus backed up his claims with an understanding of the divine character that could not be matched by the religious leaders. He taught with authority, and if the text is to be believed, Jesus reasserted his claim with physical miracles to confirm his divinity. Time and time again, scholars of the Jewish tradition tried to catch Jesus by asking him impossible questions based on the paradoxes of their faith and the dichotomy between the authority of their faith and the authority of their Roman overlords, but Jesus evaded self-incrimination by turning the enigmas back on the Pharisees. He understood the law and the divine character too well to be trapped by their interrogations.

When Jesus made the above statement, he knew full well what he was saying. He understood the timelessness of YHWH and the simultaneity of infinite presence. As an extremely wise man, he could grasp the concept of eternity. And, in as much as he might have been divine himself, he may have had an experiential understanding of complete timelessness. The theologians who came after him in his and other traditions would not have understood eternity and divinity as fully as Jesus might have, but their perspectives will be valuable for our present discussion.

In you it is not one thing to be and another to live: the supreme degree of being and the supreme degree of life are one and the same thing. You are being in a supreme degree and are immutable. In you the present day has no ending, and yet in you it has its end. (Augustine, Confessions 8)

When we conceive of the divine as *being*, it necessarily must also live. It is not one thing to *be* in the philosophical sense and to *live* in the literal. Divinity abides, completely alive and present. I AM is perhaps the only way to truly express this infinitude of being. It is all pervasive through time and space.

There is one body and one Spirit – just as you were called to the one hope that belongs to your call – one Lord, one faith, one baptism, one God and Father of all, who is over all and through all and in all. (Eph. 4:4-6)

This divine being is unlike any other life or existence in our universe. Nothing is all-pervasive, flowing over all and through all and in all. Nothing except light, perhaps: our gateway into infinity. We might glimpse this divine character, but our mortal character is of another kind.

What could be worse arrogance than the amazing madness with which I asserted myself to be by nature what you are? I was changeable and this was evident to me from the fact that I wanted to be wise and to pass from worse to better. Yet I preferred to think you mutable rather than hold that I was not what you are. (Augustine, Confessions 68)

The divine is not like we are. It is not constrained as we are. "[God's] vision of occurrences in time is not temporally conditioned" (Augustine, Confessions 221). The divine experiences and comprehends activities that happen in time, but its experience and comprehension are outside of that time. The divine experiences all in one uniform whole, understanding the totality of time and space and human will in a single instance, in every single instant.

But do not overlook this one fact, beloved, that with the Lord one day is as a thousand years, and a thousand years as one day. The

165

Lord is not slow to fulfill his promise as some count slowness, but is patient toward you, not wishing that any should perish, but that all should reach repentance. (2 Peter 3:8-9)

So too, if and when the mortals of this life attain an eternal life in the hereafter, they would be able to experience something like this eternal present. "The more perceptive of them will say, 'Your stay [on earth] was only a single day'" (Qur'an 20:104).

Divine access to eternity is not only experiential or perceptual. Infinite being must also act and speak, if it is to do anything at all, eternally. "And so by the Word coeternal with yourself, you say all that you say in simultaneity and eternity, and whatever you say will come about does come about" (Augustine, Confessions 226). The Word of God, divine wisdom, and indicative truth issue from God in a continuity of existence, and the Word itself abides in its own special existence which is at one with the existence of the divine Father.

But wisdom itself is not brought into being but is as it was and always will be. Furthermore, in this wisdom there is no past and future, but only being, since it is eternal. For to exist in the past or in the future is no property of the eternal. (Augustine, Confessions 171)

More practically, we see this eternal expression of wisdom in our mundane universe in various laws:

By [divine] law the moral customs of different regions and periods were adapted to their places and times, while that law itself remains unaltered everywhere and always. It is not one thing at one place or time, and another thing at another. (Augustine, Confessions 44)

Augustine compares these eternal laws to the laws of poetic composition, and the modern thinker can easily apply the same principles to the natural laws of physics and other sciences.

The art of poetic composition did not have different rules in different places, but had all the same rules at all times. I had not the insight to see how justice, to which good and holy people were obligated to submit, embraces within its principles all that it prescribes for all times in a far more excellent and sublime way, and, although it is in no respect subject to variation, yet it is not given all at once, but at various times it prescribed in differing contexts what is proper for the circumstances. (Augustine, Confessions 45)

For the religious thinker, the laws of the universe are embodied entirely in the divine, and thus, the divine itself is an expression of the laws it dictates. As the laws are ubiquitous and all-pervasive, so is the divine character and being, as is light itself:

I conceived even you, life of my life, as a large being, permeating infinite space on every side, penetrating the entire mass of the world, and outside this extending in all directions for immense distances without end; so earth had you, heaven had you, everything had you, and in relation to you all was finite; but you not so. Just as the sunlight meets no obstacle in the body of the air (this air which is above the earth) to stop it from passing through and penetrating it without breaking it up or splitting it, but fills it entirely: so I thought that you permeate not only the body of heaven and air and sea but even earth, and that in everything, both the greatest and smallest things, this physical frame is open to receive your presence, so that by a secret breath of life you govern all things which you created, both inwardly and outwardly. This was my conjecture, for I was incapable of thinking otherwise, but it was false... But you had not yet lightened my darkness. (Augustine, Confessions 112)

Augustine, in the above conception of the ubiquity of the divine, approaches very closely what the divine character must be like, but he himself recognized a fault with his logic, a fault that Spinoza overlooked, perhaps willfully, centuries later in his own philosophy. In conceiving the ubiquity of God as encapsulating all of the universe's physical and spiritual reality, Augustine left no space for other agents that might be alive and at work. He thus had to attribute every characteristic of the world to the divine, be it good or evil. Spinoza conceived of the evil of the world being a part of the divine character, or at least a direct result of it, and thus called 'good' what was clearly evil. Spinoza had no appreciation for agency or any substance apart from the sovereignty of God, and thus had to attribute *all* to God, resulting in the necessity of disallowing any evil in the universe, *in spite of what he perceived in the world.* He rejected what he experienced in the world 'through sheer force of intellect' to the detriment of his philosophy built on pure theory. Spinoza rejected clear reality based on his logic. Augustine, in perceiving the difficulties of understanding the divine in these terms, refined his logic to fit with that which he could verify experientially.

Lewis rectified the issues encountered by Spinoza and Augustine by recognizing that the universe as it is now has been tainted by the darkness of man's will, and only in the sanctified and glorified state of the new heaven and earth would Augustine's vision of God's ubiquity be realized. Lewis describes the heavenly scene: "The Glory flows into everyone, and back from everyone: like light and mirrors. But the light's the thing" (Lewis, The Great Divorce 343). The divine light that Augustine described did not allow for the darkness that is an unfortunate and unalterable and undeniable truth in the physical reality of our current world. He recognized this fault and modified his theology to fit with reality.

But in reality, what modern physics tells us is that light is actually all pervasive and completely ubiquitous, just not always in the visible spectrum. Vibrating atoms, possibly sub-atomic strings, release, absorb, create, and embody energy as light at all times. Light as radio waves is constantly passing through every human on earth. Light is perpetuating itself as it travels out in space, *creating* space as it moves on and on past the previous limits of our physical universe.

Again, from the perspective of Jill Bolte Taylor, neuroanatomist who experienced her own stroke:

> Everything in my visual world blended together, and with every pixel radiating energy we all flowed *en masse*, together as *one*. It was impossible for me to distinguish the boundaries between objects because everything radiated with similar energy. (Taylor 69-70)

In part, Augustine's previous quote was entirely accurate. Our physical reality, as supported by Taylor and her incredible experience, is that light pervades us all, inseparably entangling us together in an incredibly complex web of infinitesimal cause and effect built on uncertainty and probability, uniting the whole of creation together as one. The divine character and light reign sovereign. And yet Augustine knows he is wrong; uncertainty abides, probabilities wreak havoc on deterministic cause and effect, and human will refuses to submit to divine sovereignty. How can both be true?

This enigma of sovereign good and uncontrollable evil is a feature of our physical and spiritual universe, whether or not certain philosophers feel they can work it into their philosophies. We experience it. We know it. The difficulties are erased with the removal of the human element, yes; the difficulties are erased in a future glorified state; the difficulties are erased if we ignore the divine element of man; but we cannot rid ourselves of the difficulties so long as we are locked in our fleshly reality and consider that reality in total.

> All that you experience through [flesh] is only partial; you are ignorant of the whole to which the parts belong. Yet they delight you. But if your physical perception were capable of comprehending the whole and had not, for your punishment, been justly restrained to a part of the universe, you would wish everything at present in being to pass away, so that the totality of things could provide you with greater pleasure. The words we speak you hear by the same physical perception, and you have no wish that the speaker stop at each syllable. You want him to hurry on so that other syllables may come, and you may hear the whole. That is always how it is with the sum of elements out of which a unity is constituted, and the elements out of which it is constituted never exist all at the same moment. There would be more delight in all the elements than in individual pieces if only one had the capacity to perceive all of them. But far superior to these things is he who made all things, and he is our God. He does not pass away; nothing succeeds him. (Augustine, Confessions 63)

The Christian must believe in the sovereignty of God and his perfect, predestined will or else ignore significant portions of scripture. The Christian must believe in the existence of evil and man's destructive power or else ignore significant portions of scripture. The Christian must believe in the goodness of God expressed in the perfection of the world's origin and purpose or else ignore significant portions of scripture. The Christian must believe in divine judgement and righteous eternal punishment or else ignore significant portions of scripture. The Christian must believe in constituent elements of his theology that are mutually exclusive but somehow are a part of the same reality at the same time. These elements do not make sense when considered together, yet somehow when everything is perceived at the same time, the enigmas and impossibilities disappear and one glorious whole emerges. The Christian has to believe that the whole of creation is contributing to a beautiful mural that can only be appreciated in its totality. The Christian must desire for the divine poet to 'hurry on so that other syllables may come'. The Christian must desire to experience divine, eternal oneness as it is currently only experienced by the divine *one* eternally.

Italo Calvino, in *If on a Winter's Night a Traveler*, writes a character who desires just this, expressed in his imaginative ability to construct a system of mirrors that might reflect the whole universe into one perceptible image:

> Together with the centrifugal radiation that projects my image along all dimensions of space, I would like these pages also to render the opposite movement, through which I receive from the mirrors images that direct sight cannot embrace. From mirror to mirror – this is what I happen to dream of – the totality of things, the whole, the entire universe, divine wisdom could concentrate their luminous rays into a single mirror. (Calvino 166)

Any man who desires true wisdom desires this much. Scientists study the physical world in order to unearth laws that will abide. Philosophers contemplate the world of thought in an attempt to understand what ideas abide. Theologians try to overcome the obscurities of our lower world so that we might gain access to the divinity that abides. Man wishes to comprehend *all* in time and space, whether or not he believes it possible. According to scripture, the divine, in a similar way, wishes to redeem all:

> It is too light a thing that you should be my servant
>> to raise up the tribes of Jacob
>> and to bring back the preserved of Israel;
> I will make you as a light for the nations,
>> that my salvation may reach to the end of the earth. (Is. 49:6)

The divine perceives all. The divine knows all. The divine wishes to redeem all that he created. And yet we are told that there are those whom he has never known. How can it be? How can the divine power that created and sustains everything be removed from and unknown to significant portions of that creation? This is a concept dealt with in detail later, fully expounding upon the enigma; however, at present, we need only to focus on the omnipresence and omniscience and infinities and eternities of the divine.

> He has the keys to the unseen: no one knows them but Him. He knows all that is in the land and sea. No leaf falls without His knowledge, nor is there a single grain in the darkness of the earth, or anything, fresh or withered, that is not written in a clear Record. (Qur'an 6:59)

> Not even the weight of a speck of dust in the earth or sky escapes your Lord, nor anything lesser or greater: it is all written in a clear record. (Qur'an 10:61)

> Holy, holy, holy is the LORD of hosts;
> the whole earth is full of his glory! (Is. 6:3)

> But Lord what glory is there which is not in you? (Augustine, Confessions 98)

There is no monist system of belief that can conceive of its divine *one* as anything less than this. God or the divine essence must necessarily perceive and pervade all, lending its power for the existence of all else. The Jewish, Christian, and Islamic religions (not to mention many other faith traditions) believe this and complicate their beliefs by understanding God to love and care for all his creation. In his interaction with the physical world and fallen man, YHWH, God, or Allah demonstrates an odd ability to limit his eternality in order to commune with man in particular ways and at particular instances.

King Solomon, in his prayer after finishing the temple of YHWH in Jerusalem, considers this and what good the temple will actually serve:

> But will God indeed dwell with man on the earth? Behold, heaven and the highest heaven cannot contain you, how much less this house that I have built! Yet have regard to the prayer of your servant and to his plea, O LORD my God, listening to the cry and to the prayer that your servant prays before you, that your eyes may be open day and night toward this house, the place where you have promised to set your name, that you may listen to the prayer that your servant offers toward this place. (2 Chron. 6:18-20)

How can a god who created the universe be contained within that universe? How could he limit himself to one tiny location on one planet of an enormous cosmos that cannot encapsulate him in its totality? Apart from size itself, how could such a creation be able to perceive and understand the higher nature of its creator. How is it that we can conceive of this divinity at all? This begs the question, *do* we perceive this divinity in any way that approaches its reality? Perhaps we could if we were somehow glorified beyond the constraints of our current physicality.

> It is indeed most probable, that we shall then see the physical bodies of the new heaven and the new earth in such a fashion as to observe God in utter clarity and distinctness, seeing him present everywhere and governing the whole material scheme of things by means of the bodies we shall then inhabit and the bodies we shall see wherever we turn our eyes. It will not be as it is now, when the invisible realities of God are apprehended and observed through the material things of his creation, and are partially apprehended by means of a puzzling reflection in a mirror. Rather in that new age the faith, by which we believe, will have a greater reality for us than the appearance of material things which we see with our bodily eyes. Now in this present life we are in contact with fellow-beings who are alive and display the motions of life; and as soon as we see them we do not *believe* them to be alive, we *observe* the fact. We could not observe their life without their bodies; but we see it in them, without any possibility of doubt, through their bodies. Similarly, in the future life, whenever we turn the spiritual eyes of our bodies we shall discern, by means of our bodies, the incorporeal God directing the whole universe. (Augustine, City of God 1086-7)

We might have the hope of someday being able to perceive the divine directly in its essence, but in this world, we can only perceive its expressions through Platonic shadows and the blurred reflections of our imperfect physicality. Likewise, the wisdom of the divine cannot be communicated in our earthly language.

> O man, what my scripture says, I say. Yet scripture speaks in time-conditioned language, and time does not touch my Word, existing with me in equal eternity. So I see those things which through my Spirit you see, just as I also say the things which through my Spirit

you say. Accordingly, while your vision of them is temporally determined, my seeing is not temporal, just as you speak of these things in temporal terms but I do not speak in the successiveness of time. (Augustine, Confessions 300)

The language we use to describe the divine differs from divine language as much as human nature differs from divine nature. It is an imperfect, time-conditioned language that we must use. It is an imperfect, time-conditioned understanding that we must have. No wonder our words and understanding cannot contain the slippery bulk of divine enigma. No wonder we have difficulty conceptualizing the totality of our weighty reality. We lack the dimensionality that defines the divine; it is beyond us. And still, it communes with us, placing upon us the desire to experience that inaccessible dimension. This desire chafes at our consciousness and weighs on our hearts. "[YHWH] has made everything beautiful in its time. Also, he has put eternity into man's heart, yet he cannot find out what God has done from the beginning to the end" (Eccl. 3:11).

We perceive parts of the divine. Bits and pieces are within our ability to understand. But bits and pieces of God are not God. "Infinite quantity is not measurable, and cannot be composed of finite parts" (Spinoza 256). If we do not know the divine in all, we do not know the divine at all. Religions across the world sidestep this issue in ultimate, glorified knowing, in this life or the next, by which we might see the essence of the divine. We believe that we must be enlightened by the light of the divine to perceive the divine:

The man of discipline has joy,
delight, and light within;
becoming the infinite spirit,
he finds the pure calm of infinity. (Gita 5:24)

And the world is passing away along with its desires, but whoever does the will of God abides forever. (1 John 2:17)

The divine is essentially infinite. It cannot be less, elsewise it would not be divine. Man is not divine, but in him, in the will of man, his glory, the divine spark, man is allowed to commune with the divine. He catches glimpses of infinity, expressed in the impossible enigmas of this mutable and passing reality. He is given incomprehensible enigmas that express incomprehensible divinity. He is given only a little light by which he might make his way.

ENERGY AND MATTER

One possible interpretation of the significance of the relativistic unity of space and time is to regard the distinction between the two as purely man-made, a peculiarity having something to do with the nature of life and man's consciousness. Then the "velocity of light" becomes no more than a conversion factor between two sets of units, one for space and one for time, set up by man before he realized that they were one and the same thing. (March 146)

The energy-mass equivalence once again illustrates the role of the velocity of light as a conversion factor between quantities man originally regarded as distinct... This interpretation of the role of the velocity of light in the mass-energy relationship is by no means as speculative as the interpretation of its role in the space-time relationship... Both are motivated by the same spirit, a pervading one in modern physics: the desire to eliminate arbitrariness as far as possible. The idea is essentially metaphysical in nature; it expresses a faith that there are very few or perhaps ultimately no arbitrary fundamental constants in nature, such as the velocity of light. (March 156-7)

Einstein's theory of special relativity shows that light is a peculiar thing. The speed of light remains constant for every observer, regardless of what the unnatural consequences are for what they observe. One such consequence is that as objects approach the speed of light, space contracts and time passes more slowly for them compared to the rest of the physical universe. Space and time are interwoven, two dimensions of the same physical phenomenon: spacetime. Light is what remains absolute.

Of equal importance is special relativity's realization that the speed of light acts as a neat mathematical conversion factor that allows us to change immense amounts of energy into a bit of matter, or vice versa to affect massive releases of energy, as in atomic bombs. $E=mc^2$ shows us that energy can be extracted from matter and matter formed from energy.

[E=mc²] is sometimes mistakenly referred to as a formula for the conservation of energy into mass. It is more than that; *it is a statement that, for all practical purposes, the two are identical.* (March 156)

Mass and energy can be converted, one into another, not through some special process in which one is destroyed and another created, but rather because they are two expressions of the same thing, just as space and time are one and the same. Mass is conglomerated energy. Energy is displaced mass. The elements of the atom, the proton, neutron, and electron, are just other forms of the same thing that perpetuates itself through space in an electromagnetic wave. Matter is another form of light. All of it is just creative energy.

Upon finding that atoms could be destroyed into energy, and thus, that the material world is not as permanent as science had assumed, philosophers and physicists had to reconceptualize their physical conceptions. "Energy had to replace matter as what is permanent. But energy, unlike matter, is not a refinement of the common-sense notion of a 'thing'; it is merely a characteristic of physical processes" (Russell 47).

In realizing the relationship between energy and mass, the physicist stumbled upon a very unsettling fact. The world is not nearly as solid as we have always thought. We already understood that atoms contain far more empty space than physical matter, meaning that the energetic bonds that hold physical matter together compose more of the solidity of our world than does actual physical stuff. With Einstein's discovery of mass-energy equivalence, we find that even the little physical stuff in the universe is just more energy. The physical world, as we know it, can hardly even be called physical at atomic and relativistic levels.

This likely troubles the scientist and the layman alike. We may be able to understand the mathematics and the concept, but they fly in the face of our common sense and daily experience. Overwhelmed by this, man may begin to question what is real. The more religiously inclined, however, have come prepared for this eventuality. Theologians have known that this shadow-world of ours has never been as 'real' as we assumed.

In the beginning was the Word, and the Word was with God, and the Word was God. He was in the beginning with God. All things were made through him, and without him was not anything made that was made. In him was life, and the life was the light of men. The light shines in the darkness, and the darkness has not overcome it. (John 1:1-4)

Through the above quote, Christians have understood for 2,000 years what physicists are now realizing in their own unique ways. The matter of our physical world is ultimately energy. From the Judeo-Christian worldview, creation is composed entirely from the creative utterances of YHWH released in the physical world as his word, understood by Christians to be Jesus Christ himself. The character of God is composed of his will, and his will is creative in nature and imbues his creation with all sorts of life, including the pinnacle of his creation, the life of man. The darkness of the void, the absence of this creative energy and light, is shown to be hollow in the presence of the absolute physicality of the divine energy. And God's Word comes down into our mundane, imperfect, physical world to interact with his creation. The Christian understands that we are not left on earth without divine presence, with only dead words on dead pages. The Christian understands that the divine creator still reigns with active authority, "for the kingdom of God does not consist in talk but in power" (1 Cor. 4:20).

Any discussion on the beginning of the universe has always returned to a single question. Did life create stuff or did stuff create life? The secularist believes there was stuff, come from god knows where, that by a number of fortuitous events it evolved into a universe hospitable to life, and from lifeless chemicals life emerged. The religious person holds a different opinion, namely, that in the beginning there was divine life, come from god knows where, which influenced the universe to birth physical stuff and the mundane life of our world. For this discussion, we do not even have to touch on how life got to where it is now. We do not have to discuss evolution or a young earth, directed by a god or otherwise. Historically, the only question needed to consider how and why the universe *is* is whether life initially created stuff or stuff initially created life.

$E=mc^2$ asks the question in another way. Special relativity tells us that matter and physical life come from energy. Energy, that mysterious and ethereal something, is all we need at the beginning. The religious and the secularist can agree on that. No longer do we have to ask the question about life or matter, but only from whence did the energy come. Did the big bang occur at the tip of a divine finger or in the explosion of sub-atomic strings generated from some unknown intrinsic property of our universe? And what's the difference?

This question remains beyond the realm of natural human knowledge. It is an unscientific question. No tests can be done on this past event, nor could we hope to replicate it for ourselves. The philosopher can muse on possibilities and the scientist can create elaborate models, but these are little more than guesses. The theologian might assert his belief, regardless of how well established his sources are or how sure he is of his own revelation, but he has no recourse to demand that anyone else believe the same thing on grounds of logic. The question is beyond the scope of human experience and human knowledge, and thus, it is an issue of faith for everyone who holds any opinion on the answer, no matter how simple and practical.

Ancient eastern thought aligns with the idea that some intrinsic quality of the universe is what guides the physical world. Intrinsic existence could come from the divine or the purely scientific, but there is something that exists before all else. Buddhism refers to that eternal being as *mind*.

> Preceded by mind
> are phenomena,
> led by mind,
> formed by mind. (Dhammapada 1:1)

In this quote, ultimate creation itself is ignored, and the Buddha is only concerned with conservation of what already intrinsically is, namely, mind, and how it affects physical phenomena. Descartes would not necessarily disagree with this eastern approach. He himself said, "There is only a distinction of reason between conservation and creation" (Descartes 41). The effective cause of both processes must be the same, for just as I did not create life, neither can I guarantee the continuation of any existence from moment to moment, even in myself. An energy independent from my own effective energies pervades the universe and ensures the continuation of life. The energy that sustains life is much like, if not exactly the same thing as, the energy that created life. Still, this approach sheds no light on the issue of the ultimate cause of the universe.

Descartes, as many other philosophers have done, considered the nature of cause and effect, which might lend some perspective on the effect of creation. His argument is as follows:

Now, it is evident by the natural light of reason that there must be
at least as much reality in an efficient and total cause as in the effect
of that cause... It follows from this that something cannot be made
from nothing, and, likewise, that something which is more perfect
– in other words, that which contains more reality in itself – cannot
be made from that which is less perfect. (Descartes 35)

Descartes recognized the fact that for something to be created something
else must use a part of itself to affect the creation. That part of the creative
something is necessarily not the whole of the thing, and naturally the created
is less than what the creator had been, assuming the creator remains. It
certainly may be said that the creator would necessarily lose something in
order to affect the creation of something else, if only energy, and would thus
be less than what it had been. Only in the constituent elements of the
creation reconstituting into a greater whole than the sum of those elements
had been would something more perfect be affected, justifying the sacrifice
of the creator and establishing a greater order. The ultimate creative energy
that established and filled our physical universe might be considered more
perfect in itself after the act of creation and conservation only if the creation
can be considered to add to the perfection of the original energy. This is a
concept to be considered still later in this text, but the fact remains that the
creation cannot be more perfect in itself than the creative energies had been.
The physical universe cannot be more complete than the original energies
that created it.

In a completely pragmatic secularist understanding of creation,
perfection is a question of no importance. The entropy or the dilution of
creation, spreading out in our universe faster and faster, is just a cold, hard
fact. Year by year, the universe's distance from that original energy reflects
a dimmer existence, tending toward lifelessness. The original source of
energy may have created life, but only as a fortunate blip on the straight-line
path to utterly motionless death.

Suhrawardī has reason for a little more optimism than this:

Therefore, the independent incorporeal light is one. It is the Light
of Lights. Everything other than It is in need of It and has its
existence from It. It has no equal, nor any peer. It rules over all
things, and nothing rules over it or opposes it; for all sovereignty,
all power, all perfection derives from It. (Suhrawardi 87)

From this view, the ultimate source of energy in our universe is the divine light. Likewise, in the Christian understanding, the universe owes its existence and visibility to the divine light of its creator. By faith the Christian believes in him as the ultimate source of creative energy.

> Now faith is the assurance of things hoped for, the conviction of things not seen. For by it the people of old received their commendation. By faith we understand that the universe was created by the word of God, so that what is seen was not made out of things that are visible. (Heb. 11:1-3)

The physical, all the stuff of the universe that has a tangible reality, was created by intangible energies at the beginning of the universe, giving light and life to creation. The source of those energies is a matter of faith, whereas intellectual acceptance of those energies is not. Currently, there is no other explanation for the physicality of our experience in this universe apart from that original energy, converted into matter through the factor of the speed of light and creating the universe from some higher source of existence.

If one is to assume that the source of those creative energies at the beginning of creation, perhaps actualized in the big bang, is a sovereign god, there are some tempting lines of logic that one can follow. Deterministic reasoning can lead to an absolute and systematic philosophy, but one that willfully ignores the enigmas of our reality.

Benedict de Spinoza was an excellent thinker and a master of logic. His philosophy of understanding and conception of the divine are incredibly systematic and unarguably whole; however, in the assumptions he made on certain *a priori* truths, he found himself elucidating a troubling theology and heretical assumptions. In a single sentence, we can find the problems that he encountered.

> God is one, that is only one substance can be granted in the universe, and that substance is absolutely infinite. (Spinoza 255)

Spinoza carefully laid out particular definitions preceding his text in order to try to rid his philosophy of the possibility of equivocation, error, and misunderstanding. In the above quote, two definitions are of importance.

Substance – that which is in itself, and is conceived through itself

God – a being absolutely infinite – that is, a substance consisting in infinite attributes, of which each expresses eternal and infinite essentiality

Whatever Spinoza thought this God actually was, he believed that the whole of creation was composed of the essence of God, and God could only be conceptualized through the whole of creation. Following Spinoza's logic, all of creation was God, his character composed of everything good or evil or otherwise, which in Spinoza's eyes was all good because it was God, because it was his nature, because it was natural. Spinoza's religion was 'reality', and his God was natural law.

Aristotle made a similar mistake in his conception of the divine, as seen in a single quote from Bertrand Russell, summarizing Aristotle's point of view:

God exists eternally, as pure thought, happiness, complete self-fulfillment, without any unrealized purposes. The sensible world, on the contrary, is imperfect, but it has life, desire, thought of an imperfect kind, and aspiration. All living things are in a greater or less degree aware of God, and are moved to action by admiration and love of God. Thus God is the final cause of all activity. Change consists in giving form to matter, but, where sensible things are concerned, a substratum of matter always remains. Only God consists of form without matter. The world is continually evolving towards a greater degree of form, and thus becoming progressively more like God. But the process cannot be completed, because matter cannot be wholly eliminated. (Russell 169)

Spinoza felt that only through perceiving the totality of nature could we perceive God wholly. Aristotle felt that God's existence necessitated that all action and will came from and was controlled by him alone. To Aristotle, God was not 'everything', as Spinoza had assumed, but he might as well have been. Aristotle's God controlled and determined everything by his inexorable divine inspiration.

Both of these philosophers conceived of a monistic god whose character and existence left no possibility of any other wills or forces in the universe. God was everything, or, more likely, everything was God. Nature and every fact of life was merely a fact of God. No evil could exist if God was good because there was nothing else in the whole of creation apart from God and his goodness. The world *as it is* is the world *as it should be*. Nature is divine. Unfortunately, Spinoza and Aristotle must have been mistaken in this.

It would be difficult to conceive of a monist religion or philosophy in which the *one* is not sovereign. The *one* is the all-pervasive force creating and maintaining the universe, and naturally, its power will reign supreme. However, sovereignty, or *having* complete control, is not synonymous with *exercising* that control. Absolute sovereignty does not imply that there can be no other forces within the kingdom. Some of those forces may even act contradictory to the divine will. It may be possible, however, that it is a part of sovereign will to allow those forces to determine their own ends by their own means, forgoing the control that it could and otherwise would assert. It is very possible to conceive of a sovereign divinity that does not permeate the whole of creation with his inexorable will in each particular, though naturally the whole of creation would bear his divine character as the ultimate source of all physical and spiritual matter. Only through the sovereign will could mortal energies abide and attain eternality. The divine energy created everything, but that does not imply that everything is divine.

> But you, my love, for whom I faint that I may receive strength, you are not the bodies which we see, though they be up in heaven, nor even any object up there lying beyond our sight. For you have made these bodies, and you do not even hold them to be among the greatest of your creatures. How far removed you are from those fantasies of mine, fantasies of physical entities which have no existence! We have more reliable knowledge in our images of bodies which really exist, and the bodies are more certain than the images. But you are no body. Nor are you soul, which is the life of bodies; for the life of bodies is superior to bodies themselves, and a more certain object of knowledge. But you are the life of souls, the life of lives. You live in dependence only on yourself, and you never change, life of my soul. (Augustine, Confessions 42)

The Jews of the Old Testament were commanded to make no physical idols for use in shrines in their homes, or in the tabernacle or temple, or on sacred mountain tops. YHWH commanded them to make no image or likeness of him lest they grow to believe that they understood the characteristics of him. Truly, it would be impossible to accurately depict God

181

in physical form or picture, and so they were commanded to make no attempt. The Apostle Paul asserted as much in his evangelical mission in Athens, speaking to polytheists who had no moral compulsion to refrain from idol worship.

> Yet [God] is actually not far from each one of us, for
> "In him we live and move and have our being";
> as even some of your own poets have said,
> "For we are indeed his offspring."
> Being then God's offspring, we ought not to think that the divine being is like gold or silver or stone, an image formed by the art and imagination of man. (Acts 17:27-29)

Paul recognized the fact that all of creation and all of humanity must bear the likeness of God as his creation, but that we should not think that the relationship is reciprocal. It is an unequal relationship in which the divine creator lends more reality than he receives in return, as would naturally be expected, though his will is perfected in a creation that ultimately abides by that will. To man, the divine composes everything we have ever known, and all good is attributable to his power:

> Now God designed the human machine to run on Himself. He Himself is the fuel our spirits were designed to burn, or the food our spirits were designed to feed on. There is no other... God cannot give us happiness and peace apart from Himself, because it is not there. There is no such thing. (Lewis, Mere Christianity 35)

Though all material and all life were created by some ultimate energies, perhaps by something divine, neither are the same thing as those energies. Humanity may run on divine fuels, but it is not of the same character of the divine. Man's soul may have been formed in the image of God, but man's soul is not God. If there is something divine about our reality, the physicality of our earthly lives has a negative effect on the ways we might interact with that divinity. The dichotomy of our tenuous relationship with the divine can be described, "I was caught up to you by your beauty and quickly torn away from you by my weight" (Augustine, Confessions 127). Matter may just be energy coalesced into physical form; however, expressions of matter are still distinctly different from expressions of the energies that compose it. The weight of the physical differs from the effervescence of energy in the same way as fallen man differs from perfect divinity, regardless of how directly they are related.

Many theologies wrestle with the complications arising from the dichotomy of the earthly and the divine. Our banal world must be wholly unlike the glories of the heaven of heavens, the hereafter, or enlightenment. Our earthly bodies seem to contradict the divine nature, and yet in modern physics and refined theology, we find instead that the divine nature, full of enigma and mystery, may indeed be quite comfortable with the physicality of its creation.

> Christianity is almost the only one of the great religions which thoroughly approves of the body – which believes that matter is good, that God Himself once took on a human body, that some kind of body is going to be given to us even in Heaven and is going to be an essential part of our happiness, or beauty and our energy. (Lewis, Mere Christianity 59).

The above quote employs two biblical examples to confirm the assertion that Christian theology thoroughly approves of the physical in spite of the obvious beauties of the energies of the divine. Jesus' arrival on earth as a divine being shows God's willingness to commune within the confines of time and space in our universe, as also expressed in YHWH's walks with Adam and Eve in the garden of Eden. Likewise, the Bible tells its readers that our glorified existence in the afterlife will include a new perfect body through which we will experience the new creation similarly to how we experience the current creation, only more fully.

The careful reader of the Bible will encounter one more example of this divine approval of the physical world, and a knowledge of modern physics adds a depth of meaning that the ancients could not have hoped to have understood.

As Jesus enters Jerusalem on what is now known as Palm Sunday, his disciples and all those who followed him with a feverish excitement to witness the Jewish messiah were shouting praise to him. The religious leaders, for fear of their Roman oppressors or out of a concern for protecting the Jewish people from heresy, asked Jesus to rebuke these people and tell them to be quiet. Jesus responded simply by saying, "I tell you, if these were silent, the very stones would cry out" (Luke 19:40). The irony in making such a statement is that the very stones were already crying out, constantly releasing energies in the form of very long wavelengths of light, filling the universe with radio frequencies of the divine energy spiraling out of the quantum structures of the atoms which composed the stones. All of our physical universe is constantly emitting the energies which vitalize it. Insofar as those energies are divine, the physical universe is always singing the praise of that divinity.

> Your entire creation never ceases to praise you and is never silent. Every spirit continually praises you with mouth turned towards you; animals and physical matter find a voice through those who contemplate you. (Augustine, Confessions 72)

The divine energies of the big bang still pervade and compose and characterize the physical creation that resulted from the initial input at the moment of creation. The world of light and, in a lesser way, perhaps, the world of atoms, molecules, and all physical matter, bear the characteristics of the divinity that created them, whether that divinity is expressed in intrinsic laws of nature or something far more specifically imposing on human life. Creation bears the creator's character, but still, it is not quite like that creator.

> Everything God has made has some likeness to Himself. Space is like Him in its hugeness: not that the greatness of space is the same kind of greatness as God's, but it is a sort of symbol of it, or a translation of it into non-spiritual terms. Matter is like God in having energy: though, again, of course, physical energy is a different thing than the power of God. (Lewis, Mere Christianity 87)

INTRINSIC EXISTENCE

Every effect has a cause. Of course. We all know as much. But the secularist and the religiously devout both have no satisfactory answer for the 'ultimate' cause. There is absolutely no way to explain how our universe first began without also explaining what caused the cause of the universe. The human mind cannot conceive of what exactly this means and how to solve the problem.

Quantum physics offers an interesting scientific perspective on the issue. Because quantum events are based on probabilities rather than certain cause and effect, the physicist can claim that a quantum event, for which we cannot conceive a sequential cause, is what began what we now perceive as our entire universe. However, even those probabilities must rely on some underlying apparatus or state of necessity, perhaps string theory, that determines the indeterminable state. And still, an input of energy is necessary to empower the probabilities of quantum physics and string theory. Where could the energy have come from? Why is it in our universe?

In his poem *Andrea del Sarto*, Robert Browning famously penned, "Ah, but a man's reach should exceed his grasp, Or what's a heaven for?" It seems the quote applies to many of the questions we ask and the answers we seek. We continually reach out for another cause, historically finding another greater reality after each scientific discovery or paradigm shift, discontent to accept an effect without something behind it, but unable to understand how anything could be behind it at last, unless it be something profoundly and mysteriously divine.

There are particular ways we understand the universe that produce uncomfortable implications for our default perspective of Newtonian determinism. Quantum theory is difficult to unify with our commonsense understanding of cause and effect. Light betrays many of our assumptions and causes us to reconsider that which we take for granted.

For one, light seems to show us that being and existence is an intrinsic quality in our universe, a concept that could not be shed even if desired. We have already seen in the Judeo-Christian account of creation that light peopled the physical void of the universe, filling and making manifest that which otherwise was completely non-existent. In a similar manner, the character of light as understood by modern physics implies an intrinsic existence that, once existing, depends on nothing else.

> The law of the conservation of energy claims that the total energy of an isolated system cannot change, that it cannot be created or destroyed, and that mass and energy cannot be added to and are not removed from an isolated system... Upon reflection the conservation of energy law infers that all activity in the universe is self-perpetuating, which is absurd. (Pelton 23)

Absurd indeed, akin to the scientific enigmas already encountered in our treatment of the basic physics of light. Somehow the energy of the universe, as an isolated system, could not come from without and must stay within, perpetuating itself with the expansion of the universe and diluting into an incomprehensibly massive equilibrium. But if the energy that composes everything within our isolated universe could not come from without, from whence could it come at all. Somehow the existence of our universe is intrinsically tied to the existence of the light and energy that pervades every inch of it.

Light is self-perpetuating. This in itself is difficult to comprehend and counter-intuitive to the balance of physics, at least Newtonian physics. Industrialists across the world would love to find a self-perpetuating mechanism from which they could extract energy without costly inputs; however, no such apparatus is to be found. All machinery requires the input of energy to continue to function through time. Except light, that is.

So long as light encounters no force or matter that would deflect or absorb it, it will continue in a straight-line path of alternating electric and magnetic fields. Expressed as light, "electric and magnetic fields, inseparably coupled and mutually sustaining, propagate out into space as a single entity, free of charges and currents, sans matter, sans aether" (Hecht 39). Again, "bound together as a single entity, the time-varying electric and magnetic fields regenerate each other in an endless cycle" (Hecht 40). Light is a perpetual motion machine. It will continue on forever, simultaneously creating and operating in space at the speed of light beyond the previous borders of our universe, ever expanding the physical realm. Perhaps there are borders of our universe beyond which light may not travel, some intrinsic boundary that we cannot yet conceive, but physics cannot yet tell us that such borders exist. As far as we know, light will just keep going and going and going, in eternal existence.

Even in the destruction of physical matter, physics ensures via the law of the conservation of energy that this does little to destroy one iota of the creative energy that initially formed our universe. In the destruction of matter, energy is created, and in the absorption and coagulation of energies, matter is created. It is odd to find such a concept in the ancient philosophies:

> But it is ensured that the process of destruction, which results in the disappearance of mutable and mortal natures, brings what existed to non-existence in such a way as to allow the consequent production of what is destined to come into being. (Augustine, City of God 476-7)

Ancient theologies imply the same process of divine regeneration, taking that which was and forming it into a more complete and a more perfect version of what it had been:

> Of old you laid the foundation of the earth,
>   and the heavens are the work of your hands.
> They will perish, but you will remain;
>   they will all wear out like a garment.
> You will change them like a robe, and they will pass away,
>   but you are the same, and your years have no end.
>   (Ps. 102:25-7)

The source of all light, the source of all existence, has a relationship that differs from the way that the rest of creation relates to itself. The divine creative force is an intrinsic cause, completely unlike our traditional understanding of cause and effect. There is something about it that is self-actualizing and can thus act as the source of being for the rest of reality.

> Since everything other than the Light of Lights is from It, they do not depend on another in the way that some one of our actions depends on a time or the removal of an obstacle or the existence of a condition, each of which has a role in our actions. (Suhrawardi 115)

Regardless of the eternality of the sustaining energies of the universe, the lower world, created material, is not in and of itself. Even light, as mysteriously divine as it is, must have its source in another *more* self-existent thing. The physical world, even the troublingly unphysical and uncertain phenomenon of light, is something entirely different from the ultimate source of existence.

> And I considered the other things below you, and I saw that neither can they be said absolutely to be or absolutely not to be. They are because they come from you. But they are not because they are not what you are. That which truly is is that which unchangeably abides. (Augustine, Confessions 124)

The eternality and self-perpetuation of light is intrinsic to it, but the source of the energies that allow such to be the case is still difficult to conceptualize. Where did it all come from? Naturally, one must assume if there is to be an ultimate source of energy, that source would have to be completely self-perpetuating and regenerating, elsewise it would have no energies to appropriate to another entity. The life of life, the light of light, must be something that overcomes the law of the conservation of energy, sharing that which is intrinsic to it, its own life-force:

> God offers eternal life *freely* in Jesus. Yet this life is *eternal* precisely because it entails being grafted into the very life of God: becoming partakers of the divine nature, bound in the Spirit to the Son into the life of the Father. (Butler 200)

Creation must share in the energies of its creative origin if it has any hope of the conservation of its life, on earth and into the possibility of the hereafter. Energy outside of the living thing is needed to power the living thing. The light that warms us, gives physicality to our food, and shelter for our bodies, giving us life upon life, is the only energy we know of that might offer this continuation of life. It is only with this scientific, philosophical, and theological understanding of light that we can begin to appreciate that we have access to immortality through light, "immortality to light through the gospel" (2 Tim. 1:10).

PRESENT

Eternal and supreme is the infinite spirit;
its inner self is called inherent being;
its creative force, known as action,
is the source of creatures' existence. (Gita 8:3)

And that, by the way, is perhaps the most important difference
between Christianity and all other religions: that in Christianity
God is not a static thing – not even a person – but a dynamic,
pulsating activity, a life, almost a kind of drama. Almost, if you will
not think me irreverent, a kind of dance. (Lewis, Mere Christianity
95)

Light always moves at the speed of light, or it ceases to be light. The
wave nature of light places supreme importance on the moment of its
waviness: apart from present probabilities, no cause can determine which slit
a photon will pass through. A photon never ages and physical matter
travelling at the speed of light will experience no passage of time relative to
the rest of the universe. The ultimate energy of our universe must be self-
existent and eternal, experiencing an infinite present with no change to
measure or animate the passage of time. Light, an important expression of
the divine in our physical world, is eternally present, an incomprehensible
reality for us time-constrained mortals.

And although a systematic study of all theology will not at present
determine the accuracy of C.S. Lewis' quote above, Christianity, without
doubt, does espouse an understanding of the divine as being ever-present and
unquestionably active. He who would worship the god of letters found in
any religious text or the god who he wishes to affect the salvation of his
mortal life *only in the future* finds himself worshipping a divinity dead under
the weight of the past or a divinity whose activity is unreal, waiting for a
future to find life. The mature Christian believes in a God who is
omnipresent, through time and space, and who is active always, sustaining
that which he created, experiencing all activity in an eternal 'now'.

The name which the Judeo-Christian divinity uses to describe itself is an
excellent starting point for the student of light who wishes to know the divine
character. God introduces himself to Moses in these miraculous terms:

Then Moses said to God, "If I come to the people of Israel and say to them, 'The God of your fathers has sent me to you,' and they ask me, 'What is his name?' what shall I say to them?" God said to Moses, "I AM WHO I AM." And he said, "Say this to the people of Israel, 'I AM has sent me to you.'" God also said to Moses, "Say this to the people of Israel, 'The LORD, the God of your fathers, the God of Abraham, the God of Isaac, the God of Jacob, has sent me to you.' This is my name forever, and thus I am to be remembered throughout all generations.'" (Ex. 3:13-15)

The I AM in this quote is the Hebrew word commonly pronounced 'yah-weh', spelled YHWH, which has been used in this book to refer to the god of the Hebrew scriptures, the texts of the Old Testament. A discussion of the name YHWH would necessarily include much already included in this book and many other commentaries on Christian and Jewish theology besides. In addition to the synthesis of the points already made in this text and those which will soon be elucidated, one quote will suffice at present to describe what YHWH means:

I am a verb. I am that I am... I am alive, dynamic, ever active, and moving... Nouns exist because there is a created universe and physical reality, but if the universe is only a mass of nouns, it is dead. Unless 'I am,' there are no verbs, and verbs are what makes the universe alive. (Young 206)

Students of grammar will recognize the inescapability of 'I am'. This first-person present form of 'to be' is often one of the first words learned by anyone attempting a new language. The essence of existence, 'be', is applied to oneself first. Descartes proposed *Cogito Ergo Sum*, 'sum' being the Latin for the first person present 'I am'. *Being* is of utmost importance, *eternal being* is a divine quality. Yet, the present, the only reality ever known by any time-constrained entity, is a slippery idea. Augustine considered the fact that the past is gone and dead, never to *be* again. The future is unreal, waiting for its opportunity to *be*. The present is the only true expression of YHWH, and that is a horrendously incomprehensible concept.

If we can think of some bit of time which cannot be divided into even the smallest instantaneous moments, that alone is what we can call the 'present'. And this time flies so quickly from future into past that it is an interval with no duration. If it has duration, it is divisible into past and future. (Augustine, Confessions 232)

At the moment when time is passing, it can be perceived and measured. But when it has passed and is not present, it can not be. (Augustine, Confessions 233).

So indeed we cannot truly say that time exists except in the sense that it tends toward non-existence. (Augustine, Confessions 231)

Time is one of the basic tenets of our physical universe, the universe of mortal experience; however, we are told, and upon contemplation we realize, that time is not so self-evident as we once assumed.

The past is gone. Every specific observation ever made will never *be* again. We do not question the fact that it *was*, but never again will it *be*.

The future is unreal. Every future possibility is built on a mountain of unknown contingencies. Even the most effective fortune teller or predictive scientific model is constrained by the unlimited possibilities of that which not yet *is* and the waviness of the present.

Time will not be contained, primarily because the present cannot be bridled by a comprehensive, scientific understanding. It is impossibly and infinitely unknowable, although, of course, the present is all we will ever know and experience. The depths of reality seem unreal in the expression of that divine existence.

Yet time has an unconquerable effect on every moment of the mortal life. "Time is not inert. It does not roll on through our senses without affecting us" (Augustine, Confessions 60). The present, as incomprehensibly thin as it is, as impossibly real as it is, continues to assert its effect on anything conditioned within the limits of time.

No one exists for even an instant
without performing some action;
however unwilling, every being is forced
to act by the qualities of nature. (Gita 3:5)

The divine present is the foot on the gas, as previously mentioned, and every being that has any sense of existence is being propelled forward by it.

191

Activity is absolutely inescapable. Every moment activates a whirlwind of actuality.

> Although the ego center of our language center prefers defining our *self* as individual and solid, most of us are aware that we are made up of trillions of cells, gallons of water, and ultimately everything about us exists in a constant and dynamic state of activity. (Taylor 69)

The very physicality of mortal existence is hinged upon the inexorable present of activity. Everything that is *is*, and everything that is not, is not *is*. Activity defines existence.

> *Doing* itself seems not to have any volume of experience. It seems like an extensionless point, the point of a needle. This point seems to be the real agent. And the phenomenal happenings only to be consequences of this acting. "I *do*..." seems to have a definite sense, separate from all experience. (Wittgenstein, Philosophical Investigations 161e)

Philosophy, theology, and physics all agree that the present is inescapable and the absolute agent of all activity. There can be nothing else more real than the present itself. The past and future are completely immaterial and completely unreal, regardless of how much in the future or the past:

> The next hour, the next moment, is as much beyond our grasp and as much in God's care, as that a hundred years away. Care for the next minutes is just as foolish as care for the morrow, or for a day in the next thousand years – in neither can we do anything, in both God is doing everything. Those claims only of the morrow which have to be prepared to-day are of the duty of to-day; the moment which coincides with work to be done, is the moment to be minded; the next is nowhere till God has made it. (MacDonald 108)

The present is all we have. Thus, our activity in the present is the only thing that is real and the only thing to be considered as consequential. Past is gone and future is not. All that *is* defines all of our existence, all of our mortal character: "Thus, in one word, states of character arise out of like activities" (Aristotle 24). Character is defined by action, and action is defined by activity in the present, the unrelenting present. Action, in the ancient philosophical perspective, defines what is real in our physical universe, and the present is likewise influential in the modern theological perspective:

> The very act of believing in God after such a fashion that, when the time of action comes, the man will obey God, is the highest act, the deepest, loftiest righteousness of which man is capable, is at the root of all other righteousness, and the spirit of it will work till the man is perfect. (MacDonald 283)

Present obedience is the only thing that we will ever express in relation to the eternal present of the divine. Thus, as will be discussed presently, the human will is more important than can be conceived. Present action is the only reality ever known by mortal and divine alike, and present action is guided by the steering wheel of the divine aspect of human will. Nothing else expressed by mortals could be of any consequence to the divine.

> But love, in the Christian sense, does not mean an emotion. It is not a state of the feelings but of the will... Do not waste time bothering whether you 'love' your neighbor; act as if you did... When you are behaving as if you loved someone, you will presently come to love him. (Lewis, Mere Christianity 73-4)

It is said that the road to hell is paved with good intentions. If those intentions are not solidified by good activity, the statement may well be true. The intention of future good is not good. Present, active good is good. There is nothing else good besides. Thus, the present is also active in determining the future essence of that which currently is. "It may sound paradoxical, but no man is condemned for anything he has done; he is condemned for continuing to do wrong" (MacDonald 269). The reality of our mortal state is not defined by what has been done or what we plan to do, but rather, by what is done, by what we do. If it has not been sufficiently established already, the present is the only reality. Active will defines the actuality of mortal character. Nothing else contributes to our character, even the summation of all our past behaviors. The past informs. The future inspires. But only the active will abides.

If a man forget a thing, God will see to that: man is not lord of his memory or intellect. But man is lord of his will, his action; and is then verily to blame when, remembering a duty, he does not do it, but puts it off, and so forgets it. If a man lay himself out to do the immediate duty of the moment, wonderfully little forethought, I suspect, will be found needful. That forethought only is right which has to determine duty, and pass into action. To the foundation of yesterday's work well done, the work of the morrow will be sure to fit. Work done is of more consequence for the future than the foresight of an archangel. (MacDonald 109)

The totality of earthly religion emphasizes the importance of human will toward good. The human act, the mortal expression of divine good, will always be of supreme importance toward divine ends. We are told, "Be intent on action, not on the fruits of action" (Gita 2:47). The ends are of little importance if the means are disregarded. In fact, the means are the ends. Activity is totality. The present is eternally real; anything else is not.

Biblical references emphasize the same importance of the present, of the human will:

What good is it, my brothers, if someone says he has faith but does not have works?... Faith by itself, if it does not have works, is dead... Was not Abraham our father justified by works when he offered up his son Isaac on the altar? You see that faith was active along with his works, and faith was completed by his works. (James 2:14,17,21-22)

Faith is materialized by the act of will. Faith without the expression of will is nothing. Faith is only actualized by the actions which express it. Belief in the divine is expressed through the physicality and reality of the present and the totality of its action. Those who claim to believe without obedience are called liars by their own actions, which express their unbelief. Those who know not what they do, but express their obedience through divine action, are justified:

For it is not the hearers of the law who are righteous before God, but the doers of the law who will be justified. For when Gentiles, who do not have the law, by nature do what the law requires, they are a law to themselves, even though they do not have the law.

They show that the work of the law is written on their hearts, while their conscience also bears witness. (Rom 2:13-15)

But be doers of the word, and not hearers only, deceiving yourselves. For if anyone is hearer of the word and not a doer, he is like a man who looks intently at his natural face in a mirror. For he looks at himself and goes away and at once forgets what he was like. But the one who looks into the perfect law, the law of liberty, and perseveres, being no hearer who forgets but a doer who acts, he will be blessed in his doing. (James 1:22-25)

The Christian understanding of faith is intrinsically tied to the Christian understanding of obedience. Faith and works are expressed in the same eternal present, inseparably tied. Belief and action are at once the same when compared to the divinity of the present. Because nothing else *is*, no other expression of faith will ever be considered. Obedience is all.

And Christian obedience implies moral understanding, if only from the conscience, which constitutes a union with the divine unparalleled by any other potential reality. Christian obedience is compared to *being* in the eternal present of light, the light by which the divinely obedient are guided.

But you are not in the darkness, brothers, for that day to surprise you like a thief. For you are all children of light, children of the day. We are not of the night or of the darkness. So then let us not sleep, as other do, but let us keep awake and be sober. (1 Thess. 5:4-6)

Clear-minded, sober, directed, and purposeful action is what defines the life of the devout. Nothing else will express the obedience required by the divinity of our universe. Action is all that was and all that will ever be asked of us mortals. The present is all we have, and we will be judged or justified by the amalgamation of our presents, by our ultimate present. Our present decisions expressed in our present actions are naturally all that could ever define our mortal character. The will is eternally important, as is the omnipresence of the divine will.

For the word of God is living and active, sharper than any two-edged sword, piercing to the division of soul and spirit, of joints of marrow, and discerning the thoughts and intentions of the heart. (Heb. 4:12).

The word, the will of the divine, is active, as our understanding of the universe would lead us to expect. We might perceive our universe through modern physics or ancient theology, but the present is always of utmost importance, regardless of one's outlook. The present is real. The present is divine. The present is what defines the divinity within us: the human will that aligns us with any divine energies that might populate our physical world.

UNCERTAINTY & WILL

Naught happens for nothing, but everything from a ground and of necessity.
-Leucippus

Causation must start from something, and wherever it starts no cause can be assigned for the initial datum. The world may be attributed to a Creator, but even then the Creator Himself is unaccounted for. (Russell 66)

As discussed, the beginning of the universe is a topic of curious interest and one which cannot be scientifically determined. Scientific theory can lead to the probability of such and such being the initial conditions of the universe, but the event, unalterably in the past, is beyond measurement or replication. As sure as the scientist, philosopher, or theologian may be, what has passed is past. We will never have the chance to experience the truth of our origins.

However, by refining our understanding of the possibilities of what could be defined as 'the beginning', we can begin to assign probabilities and likelihood of one theory against another. Modern physics lends its perspective in recognizing the congruency of the indeterminate nature of quantum events and the indeterminacy of any ultimate cause that might have contributed to creation. We might claim that the beginning of the universe is necessarily a quantum event where the probability of an event tending toward entropy gave way to the coincidence of creation. The creation of our universe could not come from some ultimate thing, for all *things* have creators, but the universe could come from some ultimate *probability* because probabilities are intrinsic qualities without definite form. Probabilities can conceivably exist without things or cause. Probabilities offer an interesting scientific explanation for that first cause, though the detractor of this theory has ample reason to be wary of an explanation without a foundation in real energy or matter. We may be able to conceive of reality without traditional form and content, but we still must ground our universe in some real source of energy, not probability alone.

In the experiments about atomic events we have to do with things
and facts, the phenomena that are just as real as any phenomena in
daily life. *But the atoms or elementary particles themselves are not
real;* they form a world of potentialities or possibilities rather than
one of things or facts. (Rosenblum and Kuttner 130)
-Werner Heisenberg

In Heisenberg's quote, we see an impossibility of logic and an enigma of
modern physics. The experiments dealing with the quantum world acquire
real results and describe real phenomena, but the structures we understand as
the model for these phenomena are themselves unreal. The physicality of the
quantum world is a mental construct allowing the scientist to imagine causes
and effects on the quantum level. The photon may have a physical effect on
electrons and electrons may have a physical effect on a scientific apparatus,
but neither of these elementary particles have a real physical existence
beyond the effects described. The quantum world must be imagined to be
physical and solid, though in reality we know of this world only by
experimental results and mathematical theory.

It will take some time for the layman and the scientist alike to accept the
reality of this quantum enigma. All of our experience in the everyday world
implies a physicality of existence, but quantum physics, which has been
verified with unmatchable consistency, denies that anything has any real
physicality. The scientist may be able to accept this warily based on unrivaled
experimental predictability and by ignoring the logical consequences, but the
philosopher still struggles to conceptualize what this uncertainty implies
about our universe.

Uncertainty in physics arrived at a time of considerable uncertainty
among philosophers, who were splitting into camps with divergent
opinions on what the point of their own studies was. (Lindley 204)

There was a time when philosophers could make ordinary claims and
feel that the claims were completely well-founded and unquestionable, but
the uncertainty of physics sheds a bit of uncertainty on philosophical
assumptions. Aristotle once could state easily, "To entrust to chance what is
greatest and most noble would be a very defective arrangement" (Aristotle
15), but modern physics asserts with a bold confidence that the entire
universe, noble or not, has been entrusted essentially to chance, probability,

and uncertainty. The philosopher has trouble making statements of the same surety as he once could. Even the theologian, regardless of his belief in divinely-qualified truths, must pause to wrestle with the consequences of quantum theory. Modern physics causes us all to stop and reevaluate our old assumptions.

Regardless of how well other philosophies and theologies might adapt their ethoi to fit with the inescapable enigma of modern physics, Christianity once again seems uniquely prepared to deal with our new scientific realities. It will come as a surprise to many of those only superficially aware of the teachings of the Bible, but physical uncertainty and probabilities fit in well with the Christian understanding of our world's interaction with the divine.

> Strictly speaking, there are no such things as good and bad impulses. Think once again of a piano. It has not got two kinds of notes on it, the 'right' notes and the 'wrong' ones. Every single note is right at one time and wrong at another. The Moral Law is not any one instinct or set of instincts: it is something which makes a kind of tune (the tune we call goodness or right conduct) by directing the instincts. (Lewis, Mere Christianity 15)

The physicality of determinate action bears little consequence on the Christian understanding of the divine 'right', but instead, the uncertainty of will allows for a flexibility in our determinate world to align with the divine will. No single action is necessarily right at all times in all places, but instead, we find that action is only right insofar as it is guided by the will of the divine. We might say that there are probabilities that certain actions will be favorable compared to others, and as in the diffraction patterns of photons, there are some regions to be completely avoided; however, the fact remains that morality itself also lacks a certain kind of physicality. In its interaction with physical beings, it too gains particular effects, but morality is horribly slippery when we attempt to pin it down. Like a photon, morality jitters and becomes troublingly vague when we try to localize it to particular instances. Any student of Ethics 101 can say as much.

This is why it is possible for the Bible to suggest morality contradictory to our common sense. A woman can drive a tent spike through the skull of a defenseless sleeping guest and a nation might affect the genocide of an entire land, and both of these actions can be praised for their good morals. Likewise, a king can offer sacrifice to YHWH and Pharisees can pray and tithe, and God might look on them with disdain. The importance of action is troublingly tied to conscience, will, and uncertainties; human will lacks the determinacy that we had all assumed God would demand.

Without lending credence to the profundity and truth of the quantum world, C.S. Lewis penned the below quote:

> In the rest of the universe there need not be anything but the facts. Electrons and molecules behave in a certain way, and certain results follow, and that may be the whole story. But men behave in a certain way and that is not the whole story, for all the time you know that they ought to behave differently. (Lewis, Mere Christianity 18)

In a footnote, Lewis qualified this statement by saying that he did believe that there was more to the physical world. He understood that there is more to the story of the electron as there is more to the story of man. Lewis was talking about the depths of morality, and he confirmed that the world, spiritually and physically, is not as certain as we might suppose.

Henri Bergson, a philosopher of the twentieth century and a contemporary of Lewis, expressed his own understanding of the realities of uncertainty thusly, "A living being is a centre of action. It represents a certain sum of contingency entering the world, that is to say, a certain quantity of possible action" (Russell 799). The living being, the human will especially, plays a role in the macro-world of humanity similarly to how a photon contributes uncertainty in the quantum world. Some of those who espouse particular forms of predestination and those who believe in the kind of extreme philosophical determinacy elucidated in Mark Twain's *The Mysterious Stranger* would disagree with this probabilistic element of the human soul, but he who holds to such tenets must contend with the reality of our world, built on quantum probabilities, not determinate cause and effect. The universe was not determined at some past time in some past state. The universe is constantly redefining its reality in the present. Uncertainty animates the reality of the eternal present, and at the same time divine sovereignty empowers uncertainty and ensures particular ends. The classic theological concepts of free will and predestination enigmatically work hand in hand, establishing the ultimate sovereignty of divinity and the undeniable responsibility that it has laid on creation.

EFFECT AND CAUSE?

> That our actual world does not have separability is now generally
> accepted, though admitted to be a mystery. In principle, any objects
> that have ever interacted are forever entangled, and therefore what
> happens to one influences the other. (Rosenblum and Kuttner 188)

Schrödinger's Cat teaches us something very strange about the realities
of our physical world. Somehow, someway, a future observation or lack
thereof determines the results of a past action. The uncertainty of a wave
allows light to pass through neither and both slits in a double slit experiment,
but a future observation of the same phenomenon necessitates that a physical
photon passes through one slit. Until the future event occurs, the past state
remains uncertain. The definitive future solidifies a wavy past. This type of
interaction is called *entanglement*. The indeterminacy of the past state is
entangled with the probability of the future state, and until all quantum
events (an incomprehensibly huge number of interactions) are finished at the
end of time, the entanglement of all quantum events remains undetermined.
The reality of that last event, which troublingly has no future state to ground
it, actualizes the contingencies of all past possibilities. We may experience
one reality, but we must recognize, though we do not understand, that the
reality we experience is solidified by the ultimate future, the eternal present,
a divinely predestined state. Likewise, we might appreciate the realities of
the past, but somehow our present lives influence what occurred way back
when. Quantum physics turns cause and effect on its head.

The consequences of Schrödinger's thought experiment are impossibly
incomprehensible and at the same time actualize the reality we knew all
along. "Entanglement with the world *constitutes* observation" (Rosenblum
and Kuttner 151). What this quote is saying is that there is no truly isolated
system within our universe. All quantum events are entangled with all other
quantum events. There cannot be a reality in the uncertainty of Schrödinger's
superimposed states beyond the infinitesimally thin moment of the present
when the light is passing through the slit and before it interacts with any
other elementary particle. The superimposed state is completely real and
totally unreal, wholly affected by the present. So long as there is still quantum
contingency (until the end of time) uncertainty prevails; however,
entanglement implies that all uncertainty has already been resolved by our
ultimate future. The unreal state is only unreal in the present, the only reality

we ever know. The true state is only realized in a future that is entirely inaccessible to those trapped in time. This is one more complication in the amalgamation of the enigmas of light. Stated theologically:

> For every attempt to see the shape of eternity except through the lens of Time destroys our knowledge of Freedom. Witness the doctrine of Predestination which shows (truly enough) that eternal reality is not waiting for a future in which to be real; but at the price of removing Freedom which is the deeper truth of the two. And wouldn't Universalism do the same? Ye *cannot* know eternal reality by a definition. Time itself, and all acts and events that fill Time, are the definition, and it must be lived. The Lord said we were gods. How long could ye bear to look (without Time's lens) on the greatness of your own soul and the eternal reality of her choice? (Lewis, The Great Divorce 360)

The profundity of the present and the human will disallow the human intellect to totally comprehend the reality in which we live; however, the present and the human will are somehow also the determinant of what reality is. Enigmatic relationship, as we've determined scientifically, defines our universe. As man has studied nature throughout the centuries, this is not what he expected or wanted to find. Man sought harmony in nature through clear relationship of basic fact.

> It is the quest of this special classic beauty, the sense of harmony of the cosmos, which makes us *choose the facts most fitting to contribute to this harmony*. It is not the facts but the relation of things that results in the universal harmony that is the sole objective reality. (Pirsig 342)

But human scholars did not choose the facts with which we now have to contend, and throughout history, scientists have opposed the relationships we have found. Einstein himself rejected the idea that uncertainty was a cornerstone of our universe. Were it not for the objective legitimacy of quantum theory and the enigmas of light, were it not for the practical reality of human will and the inescapable present, human intellect would have rid scholarship of these difficult ideas long, long ago. We accept these truths, not because we are comfortable with them, but because they are objectively true. In spite of our discomfort, we embrace the truth. Reality is difficult.

The fullness of reality is an idea with which man continues to grapple. Energies entered, created, and sustain our universe in spite of our commonsense understanding of the law of the conservation of energy and the tendency of energy toward entropy. Energy seems to be creating more, not affecting less, as we would suppose. This perpetuation of energy aligns with our theological perspectives of God's role in nature.

> You made [the universe] not because you needed it, but from the fullness of your goodness. (Augustine, Confessions 275)

> A man's possession of goodness is in no way diminished by the arrival, or the continuance, of a sharer in it; indeed, goodness is a possession enjoyed more widely by the united affection of partners in that possession in proportion to the harmony that exists among them. (Augustine, City of God 601)

The divine creation of the physical universe could not have detracted from the goodness of that divinity, and in fact, somehow, we understand that the harmonies of creation and the perfections of a portion of humanity will actually contribute and add to divine splendor. This divine economy does not abide by any earthly laws. The fullness of ultimate energies at the beginning of time would not tend toward humanity were this not the case:

> For God would have never created a man, let alone an angel, in the foreknowledge of his future evil state, if he had not known at the same time how he would put such creatures to good use, and thus enrich the course of the world history by the kind of antithesis which gives beauty to a poem. (Augustine, City of God 449)

True, evil has a constant negative effect on that which was created by an ultimate energy or a divine goodness. The meanness of a physical world and fallen humanity are part and parcel of our universe. Dead cats and evil acts are necessitated by the glories of uncertainty and human will. But the ultimate reality affected by the totality of events perceived from the eternal present is far superior to the sum of events perceived by us time-constrained beings.

There are millions of reasons to allow pain and hurt and suffering rather than to eradicate them... but your choices are also not stronger than [God's] purposes, and [he] will use every choice you make for the ultimate good and the most loving outcome. (Young 127)

This theological understanding echoes the truths of quantum physics, as Young goes on to explain in *The Shack*:

Each choice ripples out through time and relationships, bouncing off other choices. And out of what seems to be a huge mess, [God] weaves a magnificent tapestry. (Young 178-9)

The difficulties of human will become less odious, though no less complicated, if we consider them in terms of quantum entanglement. The importance of the present and the sovereignty of totality (or eternity), which is emphasized in quantum physics, finds a home in a comprehensive theology.

And I saw that each thing is harmonious not only with its place but with its time, and that you alone are eternal and did not first begin to work after innumerable periods of time. For all periods of time both past and future neither pass away nor come except because you bring that about, and you yourself permanently abide. (Augustine, Confessions 126)

The divine experience of time realigns the mortal perspective of every action, even the existence of evil, by recognizing the relationship of all and the entanglement that contributes to the ultimate good.

For you evil does not exist at all, and not only for you but for your created universe... but in the parts of the universe, there are certain elements which are thought evil because of a conflict of interest. These elements are congruous with other elements and as such are good, and are also good in themselves. All these elements which have some mutual conflict of interest are congruous with the inferior part of the universe which we call earth... If I were to regard them in isolation, I would indeed wish for something better; but now even when they are taken alone, my duty is to praise you for them. (Augustine, Confessions 125)

The above perspective from Augustine smacks of universalism (the theology in which all existence is redeemed and sanctified for eternal life); however, little in the balance of Augustine's theology would lead us to believe he was defending such a theology. Thus, it is only possible in the above quote that Augustine is referring to the effect of God's eternal perspective of all action and the entanglement that adds to his ultimate glory. Naturally, evil does not exist for the divine because the divine good is the only reality. Evil is the counterpoint of good, the counterpoint of life and existence, thus evil is constantly contributing to and defining its own destruction. Here, it is vital to recognize the disparity between the good of creation and the essence of the divine.

> If you do not take the distinction between good and bad very seriously, then it is easy to say anything you find in this world is a part of God. But, of course, if you think some things really bad, and God really good, then you cannot talk like that. You must believe that God is separate from the world and that some of the things we see in it are contrary to His will. (Lewis, Mere Christianity 30)

> Evil men do many things contrary to the will of God; but so great is his wisdom, and so great his power, that all things which seem to oppose his will tend towards those results or ends which he himself has foreknown as good and just. (Augustine, City of God 1023)

God created man in his image. That fact will not change, and thus, with the divine ability to will, man has access to the most glorious good and the lowest, unspeakable evil comprehensible in our universe. This greatness is what necessitates the beauties of man's ability to love and his horrific ability to hate. Every mortal senses the majesty of harrowing stories of bravery that affect the salvation of other men. Likewise, we all understand the depravity of human sadism expressed in war and torture and molestation and murder. The magnitude of the human will is that it gives access to the heights and the depths of possible actions. The glory of man is in the decision to contribute actively to the divine plan of goodness. The simplicity of man is in the realization that he will only ever contribute to this expression of the sovereignty of the divine, be it through glorification or judgement.

But you use all, both those aware of it and those unaware of it, in the order which you know – and that order is just. (Augustine, Confessions 99)

We, for our part, can see no beauty in this pattern to give us delight; and the reason is that we are involved in a section of it, under our condition of mortality, and so we cannot observe the whole design, in which these small parts, which are to us so disagreeable, fit together to make a scheme of ordered beauty. (Augustine, City of God 475)

Man, in his time-conditioned state, cannot comprehend the eternal goodness of divine sovereignty, and sometimes we doubt such sovereignty is real. We are aware of the evil in our world. We see it all around. And we cannot rectify our observation of this evil to a divinity who could but will not affect the opposite results in each particular instance. But with the creation and sustenance of the human will, divinity leaves room for the decision to pursue evil. The divine will is that the mortal will would align with the divine will, but the divine also wills that mortals maintain the legitimacy of human will. The divine wills that all will aligns with that of the divine, but naturally, free will cannot be directed, lest it lose its freedom.

However, it is faulty human reasoning that supposes the existence of some particular evil detracts from the universality of sovereign good. The universe must be comprehended in its totality, in an eternal present, in order for the sovereignty of divine good to be appreciated.

It also occurs to me that when we inquire whether God's works are perfect, we should not consider some particular creature on its own but the whole universe of things. For although something may perhaps rightly seem to be very imperfect when it is considered in isolation, it is very perfect when considered as a part of the world. (Descartes 46)

That is not to say that the fallen and willfully sinful creature is perfect. Of course, such a one is defected to the point that he is in danger of divine judgement and eternal death. Yet, in the relationship of such a creature with the divine, in the divine judgement that affects the condemnation of this perverted evil, we see a greater goodness and glory. The victim of unspeakable atrocity will speak to the importance of this divine judgement in a way that those blessed with a comfortable and untroubled life can hardly appreciate. We in the west may see the evil in the world, but likely we cannot appreciate the demand for justice echoing from Rwanda and Bosnia and Germany in the wake of massive genocide.

It is difficult to appreciate man's contribution to evil in the same way that it is difficult to appreciate man's contribution to goodness. The sanctified and glorified state of man does not consist of a harp on a lonesome floating cloud. The glory of man is the communion of man and his corporate worship of the divine.

> We shall never be able, I say, to rest in the bosom of the Father, til the fatherhood is fully revealed to us in the love of the brothers. For he cannot be our father save as he is their father; and if we do not see him and feel him as their father, we cannot know him as ours. Never shall we know him aright until we rejoice and exult for our race that he is *the* Father. (MacDonald 68)

The glory of man can only be fully realized in the glory all of men. The Bible goes so far as to say that the glory of the fathers of our faith is not perfected apart from our modern faith and perfection. Quantum entanglement is on full display in such a theological concept.

> And all [the faithfully departed], though commended through their faith, did not receive what was promised, since God had provided something better for us, that apart from us they should not be made perfect. (Heb. 11:39-40)

The faith of all humankind and the sacrifice of Jesus are two elements of the total redemption of all creation. Again, many individuals within that creation will certainly be condemned for their evil will, but the totality of abiding creation must be ultimately redeemed by the divine will. In the Christian understanding, Jesus affects this redemption because, "He is the propitiation for our sins, and not for ours only but also for the sins of the whole world" (1 John 2:2). The Bible makes clear that this global propitiation does not include the salvation of each constituent element, but instead activates the redemption of all that will eternally abide, all that *is*, all that which has not already actualized its own condemnation, death, and non-existence through unrepentant sin.

The reality of the condemnation of evil and redemption of good can only be understood from the lens of the eternally present, of eternity. We must use the backwards glance of totality to appreciate the effect that good and evil will impose upon the mortal soul retrospectively.

> But ye can get some likeness of it if ye say that both good and evil, when they are full grown, become retrospective... That is what mortals misunderstand. They say of some temporal suffering, "No future bliss can make up for it," not knowing that Heaven, once attained, will work backwards and turn even that agony into a glory. And of some sinful pleasure they say "Let me have but *this* and I'll take the consequences": little dreaming how damnation will spread back and back into their past and contaminate the pleasure of sin. (Lewis, The Great Divorce 338)

The earnest Christian and the skeptical secularist will both question this perspective of Christian theology, throwing shades of doubt upon the legitimacy of such an artistic expression of the realities of the divine. But he who would dismiss this human metaphor must contend with the direct source of the Bible itself. In the following quote, pay special attention to the tense of the verbs:

> For by a single offering [Jesus] has perfected for all time those who are being sanctified. (Heb. 10:14)

This verse contains two clauses, the first from 'for' to 'those' and the second from 'who' to 'sanctified'. The first clause speaks of Jesus' death on the cross in the past tense, as it certainly was in the time of the writing of the New Testament book of Hebrews. The second clause speaks in the present tense of the modern-day saints, those whose lives were actively being changed by Christ's sacrifice. Jesus' death worked forward in time to affect the sanctification of all mortals to come. Likewise, as we saw in just a couple quotes previous to this, Hebrews also claims that the sanctification of current saints retroactively affects the glorification of the fathers of the faith. Time, from the divine perspective, is inconsequential. Action, whether past, present, or future, contributes to the fulfillment of all, of totality. The past is incomplete and uncertain without the future. The future is impossible without the past. The eternal present defines the reality of all.

With this view can we begin to appreciate the *waviness*, the lack of concrete reality, in our current age.

> Only this I know. This age of ours will one day be distant past. And the Divine Nature can change the past. Nothing is yet in its true form. (Lewis, Till We Have Faces 305)

Nothing is yet in its true form. All is superimposed. We are fully this and fully that. We are already, but not yet. These non-sensical theological truths from antiquity are realized in quantum physics. The present reality is unreal and only solidified by the totality of activity. The present is the only reality, not within the confines of time, but as perceived by the eternal present. Only in the fullness of successive time will the reality of eternity be expressed. Christians believe this incomprehensible reality is affected through the blood of Jesus Christ spilled upon a Roman cross:

> In him we have redemption through his blood, the forgiveness of our trespasses, according to the riches of his grace, which he has lavished upon us, in all wisdom and insight making known to us the mystery of his will, according to his purpose, which he set forth in Christ as a plan for the fullness of time, to unite all things in him, things in heaven and things on earth. (Eph. 1:7-10)

The divine will seems to lack sovereignty when considered alongside the particulars of evil will; however, when the divine will is considered in its proper place, in eternity, armed with the timeless agency of Christ's sacrifice and its effect on human life and physical existence, we begin to appreciate the perfection of the divine will and its influence on the physicality of our uncertain world.

Sovereign divinity interacts with the physical universe in a way that cannot be reciprocated. The eternality of the divine is unaffected by our world, yet humanity contributes to the perfection of the divine plan within the constraints of space and time. The redemption, the perfection, and the retrospective consequence of totality work throughout all time and space to affect divinely predestined intentions. Everyone who abides and is abiding will be perfected, and each one will contribute to divine righteousness, expressed either through glorification of the living or judgement of the already dead. All will be redeemed, to different effect:

I believe, to be sure, that any man who reaches Heaven will find that what he abandoned (even in the plucking out of his right eye) has not been lost: that the kernel of what he was really seeking even in his most depraved wishes will be there, beyond expectation, waiting for him in 'the High Countries'. In that sense it will be true for those who have completed the journey (and for no others) to say that good is everything and heaven everywhere. But we, at the end of this road, must not try to anticipate that retrospective vision. If we do, we are likely to embrace the false and disastrous converse and fancy that everything is good and everywhere is Heaven. (Lewis, The Great Divorce 313)

Looking through the lens of eternity, universalism makes perfect sense, *but only through the lens of eternity*. Looking through the lens of eternity, predestination makes perfect sense, *but only through the lens of eternity*. That which abides eternally will be redeemed to the divine good and will contribute to the totality of that planned perfection. That which is defined by time-conditioned evil presently affects its own non-existence by removing itself from the energy-sharing and life-giving influence of the divine good. Only when we perceive the universe eternally will we see heaven and goodness in and through all, for anything else *is* not. The divine, sovereign will finally will have accomplished its predestined intentions. In the meanwhile, wills contrary to that of the divine linger, evil lives on and affects destruction in a world groaning for redemption from the sin of man. The greatness of man is powerful enough to pervert a creation imbued with the goodness of divine energy; however, the sovereignty of the divine rectifies the perverse and removes the irredeemable, the willfully and pridefully evil. The sovereignty of the divine is the only agent in our universe powerful enough to oppose the magnitude of the human will, and in the end, nothing can or will oppose that ultimate sovereignty.

A LADDER TO THE STARS

For the *will* is the deepest, the strongest, the divinest things in man; so, I presume, is it in God, for such we find it in Jesus Christ. Here, and here only, in the relation of the two wills, God's and his own, can a man come into vital contact... with the All-in-all. When a man can and does entirely say, 'Not my will, but thine be done' –

when he so wills the will of God as to do it, then he is one with God... Thus receiving God, he becomes, in the act, a partaker of the divine nature... Obedience is but the other side of the creative will. Will is God's will, obedience is man's will; the two make one... If we do the will of God, eternal life is ours – no mere continuity of existence, for that in itself is worthless as hell, but a being that is one with the essential Life. (MacDonald 155-6)

Many readers will disagree with this quote and disagree with a visceral repulsion for the concept of free will. Regardless of the discussions visited in this text already, the idea of absolute predestination without human influence, stemming from the sovereignty of an all-powerful divine being, is so deeply entrenched in the belief system of many well-intentioned and saintly believers that one book is unlikely to dislodge the convictions held thereabouts.

However, he who holds staunchly onto the indefatigable determinism of predestination must contend with two primary arguments elucidated in detail in this current treatment of divinity and light. Firstly, the physical world as understood by quantum physics, a reality doubted by few if any practitioners of science, asserts definitively that regardless of the certainties outlined in Newtonian physics, uncertainty utterly debases arguments for absolute, physical cause-effect relationships. Though generalizations of quantum events produce the classical results expected, the summation of these events are built on the knowledge that no quantum event is entirely predictable. One aspect may be ascertained, but in our surety of that one measurement of elementary particles, another measurement becomes entirely unachievable, not just practically, but also theoretically. Cause and effect cannot be measured in a phenomenon for which certain actions are based on probabilities rather than certainties, and the scientist has no hope to rid himself of this enigma by means of more powerful forms of observation. To the core of quantum theory, cause and effect is entirely unknowable.

Why would a creator impose the reality of complete cause-effect determinism upon a race of beings composed of probabilistic energies in a world in which he otherwise bars the ability to assert any perfect relationship of cause and effect? Why would God, YHWH, or Allah create a physical man in a physical world and impose contradictory 'rules' to the two? The likelihood of this system is suspect.

Secondly, the detractor of uncertainty and free will must contend with their conception of the sovereignty of the divine. The divine exists in an eternal present, already omniscient of every past, present, and future event and choice, for no event or choice can be outside of the infinity of time that gives reality to eternity. Foreknowledge does not imply determination; foreknowledge implies the perfection of experience in eternity. Foreknowledge, from the lens of eternity, does little to repress or prohibit the expression of free will in a world conditioned by time. A sovereign divinity can practice its sovereignty by imposing the necessity of free will and the responsibility of choice (a responsibility clearly outlined in almost every sacred text), total control giving way to the possibility of evil. Even more, the existence of evil in our world necessitates such a sovereign allowance of evil if we are to believe the evils of our world are truly evil and the good of our god is truly good, as we tend to do. Elsewise, the thinker is required to conceive a good god capable of evil or a world in which the evils experienced are not actually evil. The first is logically impossible to impose on the ultimate divinity of the universe. The second betrays our common perception of genocide and rape and torture and child molestation and murders decried the world over.

The balance of this chapter assumes the possibility and necessity of the existence of divine will and also man's free will, which, unlike the divine will, can be directed toward both good and evil ends, giving rise to all trouble and pain in our mortal existence.

Many philosophers have considered the will and what such a thing must be like in our world. Descartes, an ardent Catholic philosopher, conceived of the will as the human essence; "Since the will consists in a single thing that is, as it were, indivisible, it seems as if its nature is such that nothing could be taken away from it" (Descartes 49). Descartes' will can be contrasted against human knowledge. Knowledge is developed, created, practiced, incomplete, reliant, etc. Knowledge is an effect caused by particular inputs. Will, however, requires no input at all. A baby expresses will in its cries as much as a man chooses how to respond to pain. Will is self-existent and whole as itself, though surely it can be matured and controlled within the human. Will *is* of itself, but it is also an element within man that interacts with knowledge, affectation, and ability.

Likewise, most religious believers understand the divine will to be its essence, giving form to what might otherwise be an amorphous being.

A firm light hath been set for men to look on:
    among all things that fly the mind is the swiftest.

All Gods of one accord, with one intention,
>   move unobstructed to a single purpose.
Mine ears unclose to hear, mine eye to see him;
>   the light that harbours in my spirit broadens. (Rig 6:9.5-6)

For the will and power of God is God's very self. (Augustine, Confessions 115)

In God there is Will; His will is His essence, and its principle object is the divine essence. (Russell 457)

The last quote here is Bertrand Russel's summary of Thomas Aquinas' view on the divine will. In all of these above, divinity is expressed through its willful intention given substance by total power. The essence of the divine shines through the physical world where man might experience the divine as expressed in power and will and light. The essence of man broadens upon perceiving the divine light, the very essence of divinity.

The philosopher might substitute the idea of *being* with *will* and be untroubled by any differences; however, the idea of *will* is something more than *being* alone. A chair has existence, but does not express any kind of essential will. True, the elementary particles that form the chair are behaving in 'will-like' ways (emphasizing the importance of present and the inescapability of uncertainty) but the chair only expresses these qualities in its component parts. Man, however, expresses his existence, not only through his quantum composition, but more essentially through the expression of the divinity within him, through will expressed via desire and action.

"Willing, if it is not to be a sort of wishing, must be the action itself. It cannot be allowed to stop anywhere short of the action." If it is the action, then it is so in the ordinary sense of the word; so it is speaking, writing, walking, lifting a thing, imagining something. But it is also trying, attempting, making an effort, – to speak, to write, to lift a thing, to imagine something etc. (Wittgenstein, Philosophical Investigations 160e)

Will is more than action. Will is more than desire. Will must be the combination of the two, desire manifest through action or at least the attempt to act. Will cannot stop with desire. Wishing is not will. Likewise, will cannot begin with action. Activity alone is not will. Directed and purposeful movement toward activity is will.

In his description of the heavenly spiritual state, Saint Augustine describes the perfection of this relationship between desire and action: "The one necessary condition, which meant not only going but at once arriving there, was to have the will to go" (Augustine, Confessions 147). In his redeemed state, as in his relationship with the divine, man has hope that his desire will simultaneously occur with the corresponding action. The will, in this state, cannot be confused for an impotent feeling or an accidental deed because all impediments of actualizing the will are removed.

> Nobody can always have devout feelings: and even if we could, feelings are not what God principally cares about. Christian Love, either towards God or towards man, is an affair of the will. (Lewis, Mere Christianity 74)

So too, the feelings and actions of man ought to be an excellent reflection of the invisible realities of his will; however, as in the generalizations of quantum events, there is not always such an easy relationship between cause and effect. The scientific mind who seeks to apply his own prejudices and knowledge to understanding the inner life of a fellow human risks entirely incorrect conclusions. Psychology is an imperfect science in that it cannot ever be entirely predictive, no matter how refined it becomes:

> We see only the results which a man's choices make out of his raw material. But God does not judge him on the raw material at all, but on what he has done with it… when the body dies all that will fall off him, and the real central man, the thing that chose, that made the best or the worst out of his material, will stand naked. (Lewis, Mere Christianity 55)

A man with little wealth and intelligence, a man with no power or physical self-control, a man without any inclination toward charity and with a proclivity for violence might express goodwill in small or even seemingly bad ways; however, his directed attempt to change his essence through action would be as good, if not better, than the thoughtless philanthropy of an inordinately wealthy and intelligent national politician. A righteous will might often contradict the assumptions of the worldly moralist. This is the beauty and profundity of man, delineated in *East of Eden*, as previously discussed. This is his ladder to the stars. The human will is man's ability and responsibility to act in line with divine love, regardless of his state, regardless of his past. The possession of will is a mighty power, a divinity in and of itself.

> You make a thing voluntary and then half the people do not do it. That is not what you willed, but your will has made it possible... If a thing is free to be good it is also free to be bad. (Lewis, Mere Christianity 34)

> Free will is a great good, but it was logically impossible for God to bestow free will and at the same time decree that there should be no sin. God therefore decided to make man free, although he foresaw that Adam would eat the apple, and although sin inevitably brought punishment. The world that resulted, although it contains evil, has a greater surplus of good over evil than any other possible world; it is therefore the best of all possible worlds, and the evil that it contains affords no argument against the goodness of God. (Russell 589-90)

The ends sought by the sovereign divinity of our world are clearly not focused on limiting all worldly action to completely ethical, moral, and good behavior. If this was the simple and total divine will and divinity was omnipotent, we would experience an actually good world; however, we do not see only ethical, moral, and good behavior in our world. We also experience much evil. Therefore, we cannot assume that the divine will is so simple. The ends willed by the divine must be something more.

"We call final without qualification that which is always desirable in itself and never for the sake of something else" (Aristotle 10). Aristotle believed that *happiness* was final. He believed that the form of the world was directed by this intrinsic good. The modern philosopher might accept Aristotle's definition of finality without accepting his prescriptive end in happiness.

Our end, instead, is that which has been shared with us by the divine light of our universe. Our end is *will*, desirable in itself, without qualification. Man's free will reemphasizes and does not detract from the sovereignty of the divine will. YHWH asks, "Who has first given to me, that I should repay him? Whatever is under the whole heaven is mine." (Job 41:11). Man's power comes from the divine. What man does with the power bestowed upon him by God will be his own judgement. Man's will and his moral sense are a blessing and a curse, our hope of redemption and the realization of our condemnation. It is a weighty responsibility, as we might expect from the aspect of the divine that defines our essence.

Somehow, in his use of his own free will, man is contributing to the divine activity of his creator.

> Man is not the *source* of all things, as the subjective idealists would say. Nor is he the passive observer of all things, as the objective idealists and materialists would say. The Quality which creates the world emerges as a *relationship* between man and his experience. He is a *participant* in the creation of all things. The *measure* of all things. (Pirsig 481-2)

Man's will changes the world. Man creates contingencies and realities that would otherwise not be. Man participates with the divine creator in the formation of our world. Man's misuse of will has condemned the world which now groans for the redemption from the human will to that of the divine. Man's will is his own judgement.

THE DISTORTED WILL

If will is the essence and glory of the divine and if will is man's essence and gives him access to the heights of this same glory and the hope of reunion with the divine powers that created him, then a will misemployed would ensure the condemnation of the man who wantonly wastes the divine power entrusted to him. A will contradictory to the divine will, a will perverted from its original purpose, is the ultimate and self-promoting source of evil in our universe. An evil will is the essence of evil: in the Judeo-Christian understanding, sin itself. But what is an evil will? What makes a will bad? What is sin?

> No one is doing right if he is acting against his will, even when what he is doing is good. (Augustine, Confessions 14)

This is a peculiar statement, and if it is to be fully understood, we must pause to dissect the seemingly simple sentence. Firstly, as is key to understanding Augustine's words, we have already determined that will is not alone action or desire. Will is the expression of desire through action. If a man is acting and is said to act against his will, he is acting contrary to the divine will, regardless of what he does. He does as he *wills* not. If a man desires a good thing and acts against that desire, naturally he does not act in good will or do good. Further, if a man desires a bad thing and acts against that desire, he cannot be commended for the evil he intended to do.

Man seems to have a certain role in the judgement of his own behavior. His will, not alone his desire or action, acts as judge of the essence of man. Will is at once the thing measured and that measuring. The Bible says as much by teaching men not to judge another man. Christian scripture tells us not to judge our brother in Christ for his actions, for "each one should be fully convinced in his own mind" that what he does he does for God's glory; "so then each of us will give an account of himself to God" (Rom. 14:5,12). To some degree, man is responsible for and complete in and of himself in relation to God, not in relation to his fellow man.

If the will is a self-existent thing, the yard-stick for its own judgement, independent from the balance of creation, how are we to understand what sin is?

> Where do my errors originate, then? They result from this alone: since the will extends further than the understanding, I do not restrain it within the limits of the understanding but apply it even to things I do not understand. Given that it is indifferent to those things, it is easily deflected from what is true or good and in that way I make mistaken judgements or bad choices. (Descartes 48)

Descartes believed man's ability to desire and act extended beyond his ability to reason, which may well be true; however, a Biblical reading conceptualizes sin in the face of understanding. We know what to do and yet do not do it. An evil will arises from more than just a lack of understanding or even an evil desire, though evil acts may well originate from such sources. The evil will is a self-existent, independent thing. Nothing affects it. Nothing creates it. It just is.

> For nothing causes an evil will, since it is the evil will itself which causes the evil act; and that means that the evil choice is the efficient cause of an evil act, whereas there is no efficient cause of an evil choice; since if anything exists, it either has, or has not, a will. If it has, that will is either good or bad; and if it is good, will

anyone be fool enough to say that a good will causes an evil will? (Augustine, City of God 477)

The divine will could not create an evil will, yet the human will, once created and independent of its source, is able to make an evil choice, affect an evil act, and thus, create evil in an otherwise good world. The evil will has no cause. Though the will is empowered by the divine, the evil will just is. God, if divine, has no hand in evil.

> God does not guide the mischief of the treacherous. I do not pretend
> to be blameless, for man's very soul incites him to evil unless my
> Lord shows mercy. (Qur'an 12:52-3)

All beings with will tend away from the divine will because of the perfection of the divine and the natural differences between the creation and creator. Man and his will tend toward evil without the grace and mercy of the divine, which is showered upon all and invigorated by a realignment of the human will toward its predestined, divine goodness. The essence of man divides him into one of two camps, damned in himself or redeemed through the divine, already dead or eternally living, but the direction of his essence is not predetermined. Will is self-determined:

> The contrasted aims of the good and the evil angels did not arise
> from any difference in nature or origin... We must believe that the
> difference had its origin in their wills and desires. (Augustine, City
> of God 471)

So, the essence of man, and perhaps angels as well, is not forced via cause and effect toward good or evil, but regardless, compared to the perfection of the divine, man is constantly measured short of the eternal benchmarks of righteousness. We have already determined that man's will acts as its own form and source of judgement. However, even positive behavior will be short of the divine perfection. Man, though he might be right in his own eyes, and though his own eyes be objective and reasonable, will constantly find himself falling short of the divine standard, unable to affect his own salvation and continuation of life.

> I hold fast to my righteousness and will not let it go;
>   my heart does not reproach me for any of my days. (Job 27:6)

Truly I know that [God favors the righteous].
>But how can a man be in the right before God?
If one wished to contend with him,
>one could not answer him once in a thousand times.
He is wise in heart and mighty in strength
>—who has hardened himself against him, and succeeded?...
Though I am in the right, I cannot answer him;
>I must appeal for mercy to my judge...
Though I am in the right, my own mouth would condemn me;
>though I am blameless, he would prove me perverse...
For he is not a man, as I am, that I might answer him,
>that we should come to trial together.
There is no arbiter between us,
>who might lay his hand on us both. (Job 9:2-4,15,20,32-33)

In his story, Job grappled with the truth of his suffering. He knew he had committed no wrong according to the law and his own conscience, yet it verily seemed that he was being punished. His earthly losses were almost too much to bear. Job would not lose grip of his self-righteousness. His self-existent means of judgement showed that he had not willed evil; however, his existence as man as opposed to the divinity of God condemned him as imperfect. Job found that his condemnation did not lie alone in an unquestionably evil will, but rather, he found that a will misaligned with God's, as all must be (our essence is like but not equal to the divine essence), was enough for temporal pain and eternal condemnation. Evil is not a thing in itself that one can point to and pick up and analyze as an independent entity; it is a perversion of a good thing. Evil is hardly real in itself, and it has no being apart from the divine existence of will.

> I inquired what wickedness is; and I did not find a substance but a perversity of will twisted away from the highest substance, you O God, towards inferior things, rejecting its own inner life and swelling with external matter. (Augustine, Confessions 126)

> For whatever does not proceed from faith is sin. (Rom. 14:23)

Human will is human essence. The divine power, grafted into our being, is what grants us any existence at all. Divine energies created our world, and the divine essence imbued man with a higher reality of being. The divine will is the divine essence, and man perverts his own essence away from the life-giving influence of the divine. He brings about an effect in our world contrary to the intentions of the divine. Man is the source of evil in this world, the source of everything contrary to the goodness of the divine.

> The breath of life, which gives life to everything, and is the creator of every body and every created spirit (breath), is God himself, the uncreated spirit... Just as he is the creator of all natures, so he is the giver of all power of achievement, but not of all acts of will. Evil wills do not proceed from him because they are contrary to the nature which proceeds from him. (Augustine, City of God 193)

> The power of hell is rooted in our distorted desires – the wicked root, the wildfire's spark, the poisoned well – or, in classical language, our corrupted will. (Butler 59)

> Hell gains entrance into God's good world through us. *We* are the agents of destruction, the architects of demolition. God is not the architect of hell, the creator of its soul-destroying power; we are. (Butler 24)

The will of man, the distortion of the divine good, is what affects and creates and empowers evil in our world. The divine is good and omnipotent, yet divinity created other beings of will in the world, and thus, the power of our wills, created by and sustained by divine power, have the ability of evil. Sin, the distorted will, is sustained by the power of the divine, God's will toward free-will. Sin enters the world through our partial and incomplete god-likeness, our responsibility to be good, to exist as perfect goodness, and our inability to do so. Divinity creates the immeasurable powers of will, leaving the door open to evil, in the intention of also allowing the incomprehensible glories of mortal goodness and sanctification. The glorification of man affects a spectacular reality of unity between God and man; however, this reality comes at the cost of God allowing, not causing, the depths of evil of which man is capable.

> Let no one say when he is tempted, "I am being tempted by God," for God cannot be tempted with evil, and he himself tempts no one. But each person is tempted when he is lured and enticed by his own

desire. The desire when it has conceived gives birth to sin, and sin when it is fully grown brings forth death. (James 1:13-15)

The divine will, of course, cannot will (or tempt) other wills to act contrary to the divine will, nor can the essence of life affect death. This is a logical and practical impossibility. Naturally, then, another self-determining spirit in the universe, the human will, the light of man, pulls us away from the will of the divine, leading to contrary desire which leads to sin which leads to death, ultimately separating man from the life-giving influence of the divine. Man affects his own tendency toward non-existence. Man affected his own death in the garden of Eden, and now he cannot abate the perpetuation of the unfortunate evils that characterize our broken world.

According to Stoic philosophers of ancient Greek culture, the divine must be an essential element of the physical and created world. Modern philosophers, if they allow for an all-pervasive and universal law, must propose the same thing:

> God is not separate from the world; He is the soul of the world, and each of us contains a part of the Divine Fire. All things are parts of one single system, which is called Nature; the individual life is good when it is in harmony with Nature. In one sense, *every* life is in harmony with Nature, since it is such as Nature's laws have caused it to be; but in another sense a human life is only in harmony with Nature when the individual will is directed to ends which are among those of Nature. *Virtue* consists in a *will* which is in agreement with Nature. (Russell 254)

In some sense, human life is independent of the divine, and the human will is self-existent. In another sense, human life (as all creation) is dependent on the life-giving power of the divine, and the human will only gains its power through the self-existent and regenerate nature of the divine. This paradox is akin to all scientific paradoxes concerning light. All will is one, yet all wills are independent and self-existent.

> Do not imagine that the incorporeal lights become a single thing after separation from the body, for two things do not become one... The incorporeals do not cease to be, for they are distinguished intelligibly through their cognizance of themselves, through their

cognizance of their lights and the illuminations of their lights. (Suhrawardi 148)

Somehow the power of the divine will becomes an independent entity when bestowed from the source of light and life; however, independent lights cannot abide apart from the life-giving source of their independent powers. Will is self-existent, and yet somehow dependent on the ultimate source of life.

> From him come all powers, but not all wills. What they mean by 'destiny' is principally the will of the supreme God, whose power extends invincibly through all things. (Augustine, City of God 189)

The power to will comes from a divine source, but the direction of will is self-existent. If this concept is not obvious to the reader, she or he can hardly be blamed. The enigma is incomprehensible. Yet, unquestionably, the will is powered by a divine source, though it is independently able to pursue its own course. Divinity powers good and evil, in that it powers the ability to will, which is free and self-existent, if it is truly will. The present has its divine foot on the gas; the will controls the steering wheel. Without a forward motion, the direction of the steering wheel means little. Without life and action, will loses the nature of its self-existent reality.

> [The will] cannot act from itself, save in God; acting from what seems itself without God, is no action at all, it is a mere yielding to impulse. All within is disorder and spasm. (MacDonald 116)

Without the influence of the divine, the turnings of the evil will are nothing but the aimless movements of a powerless perversion of reality. Herein, the effects of divine desire are inconsequential, though the effects on the will directed toward hell are incomparable: "The consequence of a distorted will is passion. By servitude to passion, habit is formed, and habit to which there is no resistance becomes necessity" (Augustine, Confessions 140). The evil will forms and directs itself into total depravity, unable to stop the necessity of its own destruction. In its direction toward evil, the depraved will is forced forth by the divinely prompted present, a relentless activity that compels the will toward one end or another. That which began as an inclination quickly becomes an essence; character becomes slavery:

'To begin with the state, is a state ruled by a tyrant in a condition of freedom or slavery.'

'It is in complete slavery.'...

'If the individual is analogous to the state, he must be similarly placed. His mind will be burdened with servile restrictions, because the best elements in him will be enslaved and completely controlled by a minority of the lowest and most lunatic impulses... So the mind in which there is a tyranny will also be least able to do what, as a whole, it wishes, because it is under the compulsive drive of madness, and so full of confusion and remorse.' (Plato 316)

Plato, whose work came centuries before Christ, preempted the sentiments of the Apostle Paul, who discussed at length the servitude perpetuated by sinful inclinations. Though man is empowered and directed by a good and divine source, his actions might be characterized by an unexpected godlessness. His will is his own, as is his evil, condemning him through his own essence.

The curious thing, as illuminated in Plato's allegorical discussion of the nation state, is that the more a self-existent entity has power and proclivity to submit to the divine direction of its source, the freer it is. In other words, the more inclined a thing is to obey, the freer it is to accept its own direction.

This concept seems contradictory; however, upon contemplation of the evil will, the freedom of choice is only realized in the freedom toward goodness:

> Nor is it true that, in order to be free, I must be capable of moving in either direction; on the contrary, the more I am inclined in one direction the more freely I choose it, either because I clearly recognize it as being true and good or because God so disposes my innermost thoughts. Surely neither divine grace nor natural knowledge ever diminishes freedom; instead, they increase and strengthen it. (Descartes 47)

To the secularist, this concept may seem entirely unacceptable and illogical. To him, free choice implies impartial choice: the undirected desire of the will directs action whichever way it leans. However, to the religiously inclined, this freedom makes total sense. The desire of a free will must favor preferable goods rather than evils which are to be avoided. The free will should be freed of its own natural and servile tendencies, only after which it might favor the freedom of goodness and righteousness which lead toward felicity, Aristotle's happiness. Freedom to choose evil can lead to evil, and though it is freedom, no man would choose such a path had he the forethought to avoid it, thus it is a form of slavery to the evil of our will.

> For the trouble is that one part of you is on [God's] side and really agrees with his disapproval of human greed and trickery and exploitation. You may want Him to make an exception in your own case, to let you off this one time; but you know at bottom that unless the power behind the world really and unalterably detests that sort of behavior, then He cannot be good. (Lewis, Mere Christianity 24)

Our intuition, which allows us to understand the innate qualities of good and evil, allows us to understand our own innate qualities of good and evil. We believe our good to be a blessing and our bad to be a curse, and we hope that the divinity of the universe will excuse our particular evils while condemning the rest, so that we might gain access to a perfect, eternal existence, in spite of our own imperfection. All other evil must be removed lest eternity is not an unquestionable good. We know that we deserve the condemnation that comes, part and parcel, with our sin, but we hope that somehow we might be pulled out of our own servitude to sin into the freedom and glorious light of righteousness.

> God is light, and in him is no darkness at all. If we say we have fellowship with him while we walk in darkness, we lie and do not practice the truth. But if we walk in the light, as he is in the light, we have fellowship with one another, and the blood of Jesus his Son cleanses us from all sin. (1 John 1:5-7)

If not expressed clearly in all the arguments and quotations to this point, good is from the divine and evil is from a distorted will, only made possible through human will. "Anything good that happens to you is from God; anything bad is from yourself" (Qur'an 4:79). Differing religions can typically agree on this much. God is divine. The divine is good. Man perverts goodness. Sin is the source of all evil. Sin is contradictory to the divine will and removes us from communion with it. We see as much in Eden and in every sin since. We might hope that we can gain an existence apart from the divine will, we might hope to assert our own will over that of others;

however, "claims to mastery can only breed strife" (Russell 790). Claiming independence only affects our own slavery and death.

THE DIVINE WILL

The will is an essential thing in our universe. It is indivisible. It only ever operates in the present. Its existence is immeasurable.

As the will is essential and indivisible, naturally, its source must also be divine in the same sense, but so much more so, in order that the ultimate cause might affect the creation of such an essence. The divine will is a magnificent and glorious thing, and the only way a mortal might perceive such a thing is in its relationship to our mortal world, for no other lens is available. The divine will must be contrasted against the human will for us to gain some comprehension of what it may be and what it is not.

One clue to the utter difference between the human will and the divine will is through the biblical account of the existence of Jesus Christ. In the Christian understanding, Jesus was a mortal man who lived as the intermediary between God and man, sharing the very existence of God himself in mortal form. In theological verbiage, Jesus was fully man and fully God, an enigma of superimposition that cannot be fully comprehended.

As previously discussed, the will of God is said to compose his being: God's will is his essence. We cannot understand what God might be without his will. Interestingly, it appears that Jesus, though he is said to have shared the divine essence, did not share its will entirely, as we would suppose.

Jesus had his own will, which allowed him, in the garden of Gethsemane, to pray, "Father, if you are willing, remove this cup from me. Nevertheless, not my will, but yours, be done" (Luke 22:42). Jesus understood that his desires might not align perfectly with the plans of the father, but Jesus also shows us in this prayer and his following actions what sharing the divine will and eternal life means. As we have said, will is not only composed of a certain powerless desire. Will is actualized in the present through action. Though his temporal desires were not necessarily the same as those of God the Father, Jesus' will was exactly the same as the divine will, for Jesus acted constantly in alignment with the desires of his father. Jesus' actions personified the will of God in the created world. Jesus gave mortal life to the divine will by practicing the human element of the divine will: namely, obedience. Jesus acted as a mediator between God and man by showing man that to share in the immortal divinity of the Father, we must desire and pursue obedience.

Our belief in the divine is meaningless if we do not express faith in its goodness by behaving well. And as we have determined in our discussion of will, right action is only incidental goodness without a desire to live in accord with the divine will. Desire and action together, desire for and obedience to divine intentions, give man access to eternally predestined, personal good.

We all feel a certain inclination toward divinity, a desire for greatness:

> What is the light which shines right through me and strikes my heart without hurting? It fills me with a terror and burning love: with terror inasmuch as I am utterly other than it, with burning love in that I am akin to it. (Augustine, Confessions 227)

Though all men might feel the void of an existence yet unperfected by the divine light, individuals tend not to realize what it takes to accept that universal life-giving influence into their own lives. The eastern traditions take us halfway to that glory.

> Know that discipline, Arjuna,
> is what men call renunciation;
> no man is disciplined
> without renouncing willful intent. (Gita 6:2)

Hinduism and Buddhism teach us that we must rid ourselves of earthly desire in order to attain the glorification of enlightenment; however, an understanding of light would necessitate that something else take the place of mortal desire, lest an immaterial void abides. A divine desire expressed through obedience is what should replace mortal longings. Attaining the divine will becomes the primary goal of all human life. "Thus a good servant would regard the will of God as his great resource, and he would be enriched in his mind by close attendance on God's will" (Augustine, City of God 18). King David of Israel, one of the great fathers of the Jewish and Christian faiths, shares this wisdom with his son Solomon: "Know the God of your father and serve him with a whole heart and with a willing mind, for the LORD searches all hearts and understands every plan and thought" (1 Chron. 28:9). Man must know God to abide by his will.

226

To do a good thing is to do a good thing; to know God is to be good. It is not to make us do all things right he cares, but to make us hunger and thirst after a righteous possessing [by] which we shall never need to think of what is good or is not good, but shall refuse the evil and choose the good by a motion of the will which is at once necessity and choice. (MacDonald 91)

The will of the divine becomes the ends and the means. By it we are guided to the happy and free necessity of obedience, and to it we are drawn into the eternal and superior life that might be accessible through the divine. He who intimately knows the divine powers of the universe attends to their will and seeks the direction they might provide: "Send out your light and your truth; let them lead me" (Ps. 43:3). Again, the mystery of Jesus' divinity and his unity of will with the Father express the directive role of the divine will. Jesus said, "I can do nothing on my own. As I hear, I judge, and my judgement is just, because I seek not my own will but the will of him who sent me" (John 5:30). Jesus sets the example of mortal man seeking, obeying, and living the will of God, and likewise, Christianity holds that God greatly desires that man would do so. Jesus' attendance to the will of God is why the divinity of the universe is said to be well pleased with his son. The Christian God wants to reinstitute unadulterated communion with his creation, that which he purposed in the garden of Eden.

We want to share with you the love and joy and freedom and light that we already know within ourselves. We created you, the human, to be in face-to-face relationship with us, to join our circle of love. As difficult as it will be for you to understand, everything that has taken place is occurring exactly according to this purpose, without violating choice or will. (Young 126-7)

[God] desires all people to be saved and to come to the knowledge of the truth. (1 Tim. 2:4)

This is not a refrain unfamiliar to religions around the world. Man, so long as he is aware of his imperfections and yet desires the glorification of eternal life, must conceive of a deity that allows for error and works to correct such error in man. In the Rig Veda, the Hindu prays to his deities, "Now, O Ādityas, grant to us the shelter that lets man go free, Yea, even the sinner from his sin, ye Bounteous Gods" (Rig 8:18.12). Freedom from the distorted

227

will of man, from sin itself, is a protection that could be provided by the divine, a protection sought by immature believers any faith. Children seek the divine influence to keep them from sinning, an influence unavailable to them in full. Practical experience tells us that we always have the ability and option to sin. The maturing believer, however, accepts responsibility for her own actions while still hoping for divine provision of self-control.

There is a certain duality of responsibility split between the mortal will which labors under its own inadequacy and the divine will which delights to help those who seek support. The duality expresses action from both agents: "Work out your own salvation with fear and trembling, for it is God who works in you, both to will and to work for his good pleasure" (Phil. 2:12-3). The mortal is ultimately responsible for her own behavior, yet the divine is ultimately responsible for the grace that allows the mortal to attain glorification and perfection. This dichotomy operates through the enigma of the will.

> There are only two kinds of people in the end: those who say to God, "Thy will be done," and those to whom God says, in the end, "*Thy* will be done." All that are in Hell, choose it. Without that self-choice there could be no Hell. No soul that seriously and constantly desires joy will ever miss it. Those who seek find. To those who knock it is opened. (Lewis, The Great Divorce 340)

That is not to say that 'no soul that seriously and constantly desires perfect behavior will ever miss it'. Every man, woman, and child knows that this is a practical impossibility, and the theoretical principles of uncertainty tell us that no created thing operates on such simple premises. The will is far too 'wave-like' to be so consistent. And a divine creator who produced such physical creatures would not expect perfection in our current physical state.

> Now, once again, what God cares about is not exactly our actions. What He cares about is that we should be creatures of a certain kind or quality – the kind of creatures He intended us to be – creatures related to Himself in a certain way. (Lewis, Mere Christianity 80)

The relation referred to here can be nothing but a relation of the will, the mixture of desire and action requiring a supernatural understanding of the man to determine the quality of the man. No written psychological or behavioral report would capture the reality of even the basest human will. A divine interpretation is necessary:

> I the LORD search the heart
>> and test the mind,
> to give every man according to his way,
>> according to the fruit of his deeds. (Jer. 17:10)

The pragmatic thinker might question why YHWH needs to search the heart and mind of man to determine how to award man for his deeds, his behaviors, his way. Isn't man's behavior enough evidence for or against him? There is no simple explanation for the enigma of faith and works. Somehow both acts and belief are independently and cooperatively final in testing the goodness of a man. The enigma is profound, and so too we cannot understand the operation of the divine in affecting either or both of these spiritual indicators of will.

> If you say There can be but one perfect way, I answer, Yet the perfect way to bring a thing so far, to a certain crisis, can ill be the perfect way to carry it on after that crisis: the plan will have to change then. And as this crisis depends on a will, all cannot be in exact, though in live preparation for it. We must remember that God is not occupied with a grand toy of worlds and suns and planets, of attractions and repulsions, of agglomerations and crystallizations, of forces and waves; that these constitute a portion of his workshops and tools for the bringing out of righteous men and women to fill his house of love withal. (MacDonald 123)

The Christian God and any creator of our world of enigma, uncertainty, and wavy existence, is not concerned with the cold facts or physical realities of our lives. Rather, he is more concerned with the will, the ever-present expression of obedience in our ever-present existence. Man can never hold onto the past nor grasp the future, but the present is our constant companion, in which our will has the profoundest effect. Neither desire nor regret, neither past behavior nor future actions affect our reality. The present is the only lens through which our lives can properly be viewed, and the will is the only human reality of the present. If our present will is the only reality accessible to us, we ought to ensure, as much as possible, that it aligns with

229

the eternally active and operative divine will, the source and guiding light of our lives.

The divine will, especially in its interactions with and expressions in the physical and mortal realm, not only generates and sustains the life of all else, but it also is a powerful standard by which the human will might be tested. If divinity was not so intent on communing with the universe, we might have no recognition of our perilous state, but communicated in the law, morality, conscience, and a heavenly messiah, that which affects our glory and redemption is the same means by which we are condemned. The human will and the divine will share this two-fold purpose.

> Behold, the days are coming, declares the LORD, when I will raise up for David a righteous Branch, and he shall reign as king and deal wisely, and shall execute justice and righteousness in the land. In his days Judah will be saved, and Israel will dwell securely. And this is the name by which he will be called: 'The LORD is our righteousness.' (Jer. 23:5-6)

Constantly in the prophecies of Judaism, Christianity, and Islam, the god who will bring about the salvation of those who remain steadfastly dedicated to him must necessarily also affect the justice he desires, a justice attainable only through judgement of those wills contrary to his own. Many religious folks apply God's redemption exclusively to those who believe in a single, narrow, and particular interpretation of scripture, and they reserve the judgement for all those who disagree. The Christian must understand this truth broadly, perhaps more broadly than many Christians will find comfortable; however, a close reading of the Bible will reveal that, "We do know that no man can be saved except through Christ; we do not know that only those who know Him can be saved through Him" (Lewis, Mere Christianity 41-2). Jesus reveals through a parable that the self-righteous will be astounded at the judgement God has planned for them, and many who did not know who it was they served will find themselves favored by God. Judgement is not a process understood by the mortal or through morality. The active will of the divine perceives far more about the human will than we know about ourselves, and his judgement will be based on those sober observations.

God opposes the proud, but gives grace to the humble. (James 4:6)

And he who searches hearts knows what is the mind of the Spirit, because the Spirit intercedes for the saints according to the will of God. (Rom. 8:27)

For those who do practice humility and allow the divine to search their hearts, they will realize that they not only have access to divine mercy at the end of their days, but they will also enjoy the benefits of the divine grace that proactively keeps them from the evil they might otherwise do. YHWH and God promise to help shelter his beloved from the effects of others' evil wills as well as from the perversion of their own will toward evil. The divine will saves his humble followers from incidental sins. YHWH told Abimelech, an ancient king, "Yes, I know that you have done this in the integrity of your heart, and it was I who kept you from sinning against me" (Gen 20:6). Abimelech would have accidentally sinned against YHWH and man, causing sin and strife, but for the divine grace that kept him from sinning. Because YHWH knew the heart of the man, his intentions and desires, he barred the king from sinful action. But without that will to righteousness in Abimelech, there would have been no resource for him to rely on to keep him from accidentally sinning.

Many people question why divine grace might be bestowed on some and withheld from others, but Job, a supremely self-righteous man in the Old Testament, reluctantly came to realize his own worthlessness compared to the divinity of the creator.

Therefore, hear me, you men of understanding:
>far be it from God that he should do wickedness,
>and from the Almighty that he should do wrong.
For according to the work of a man he will repay him,
>and according to his ways he will make it befall him.
Of a truth, God will not do wickedly,
>and the almighty will not pervert justice.
Who gave him charge over the earth,
>and who laid on him the whole world? (Job 34:10-13)

The creator of the universe established the energies that formed, sustain, and permeate his creation. The creator's energy and will are what define his creation: they are natural, they are good. The creator of everything, the creator the laws of our physical world, also establishes the righteousness and goodness of that world, the moral law. God is good. The divine is supreme. There is no other way to conceptualize an all-powerful divinity.

Thus, any behavior of man contrary to the nature and will of the divine will result in an imperfection and a consequence. The natural result of sin is injury, for any deviation from life affects some sort of death.

> Your punishment is that which human beings do to their own injury because, even when they are sinning against you, their wicked actions are against their own souls. Iniquity lies to itself, when men either corrupt or pervert their own nature which you made and ordered, or when people immoderately use what is allowed, or when, turning to what is forbidden, they indulge a burning lust for that use which is contrary to nature. (Augustine, Confessions 47)

Man's perversion of the perfect will is his own judgement, and more, his own punishment. Man is the world's source of evil. Man is his own source of pain. Man is also a terrible source of pain for his fellow man. Man condemned the entirety of our physical universe, perhaps through quantum entanglement, and now all must be redeemed to its original perfection desired by the perfect will of the divine.

Remarkably, just as man affects his own judgement, his sanctified will and the story of his redeemed life can work to complete his salvation. We are told the salvation of man is a joint effort between man and divinity against the accuser of our souls, the prosecuting attorney seeking our destruction, Satan and our own damning wills. "They have conquered him by the blood of the Lamb and by the word of their testimony" (Rev. 12:11). Man's testimony, the story of his redeemed will, is a witness of his salvation, just as man's perverted will is a witness of his utter depravity. The human will acts alongside the divine will as both judge and savior. The human will is given this eternally significant power by the divine will.

Were it not for the intention of the divine to commune with man, man would have little reason to fear judgement or the power of his own sins.

> If I had not come and spoken to them, they would not have been guilty of sin, but now they have no excuse for their sin. (John 15:22)

> And this is the judgement: the light has come into the world, and people loved the darkness rather than the light because their works

were evil. For everyone who does wicked things hates the light and does not come to the light, lest his works should be exposed. But whoever does what is true comes to the light, so that it may be clearly seen that his works have been carried out in God. (John 3:19-21)

For if we go on sinning deliberately after receiving the knowledge of the truth, there no longer remains a sacrifice for sins, but a fearful expectation of judgement. (Heb. 10:26-7)

The pragmatic secularist and any who has difficulty in accepting the theology of judgement will say, 'How can a good god condemn anything in his creation to eternal damnation? Why would he purposefully create to only later destroy?' Giants of many faiths have rejected the consequences of this question and accepted the theology of universalism. They reject the judgement of the divine and the fiery heat of its light.

However, in rejecting the judgement of God, one has to either ignore divine goodness or its communion with man. If God is good, he must judge evil and remove it, disallowing its admittance into his new, perfect creation. If the will of man is a divine thing and God will not betray man's nature as an independent being, then God must judge and remove evil men from his divine life.

Additionally, if God is to share the goodness of his divine life with man, the utter reality of that light will make the depravity of the evil will inescapably evident. And without the goodness of the light, man would not experience the goodness of God. Either man's will, his divine element, is unreal or God's goodness has to be unreal in order for universalism to be logical. There is no other possibility. And as we have determined, the goodness of the divine is undeniable and the glory of man is experienced by us all and given ample support in many holy scriptures and well-respected philosophies and sciences. Universalism is impossible in light of a good god and human will. Since we have established both of these realities in great depth, universalism must go, and we will have to consider the uncomfortable consequences of judgement.

The presence of YHWH, to those whom oppose him, comes in as a dark day. A fearful darkness will abide for those who would turn away from the life-giving light of the divine. Sinners orchestrate their own judgement, condemnation, and damnation through their misapplication of the power and responsibility of the will, and the presence of God convicts those who have been awaiting the eternal effects of the divine will. The gloom is palpable for these. Judaic, Christian, and Muslim sources all agree on the darkness of judgement. Contrary to what we have come to expect from the brilliance of the divine, holy displeasure is ultimate darkness, formless void.

> Behold, the day of the LORD comes,
>> cruel, with wrath and fierce anger,
> to make the land a desolation
>> and to destroy its sinner from it.
> For the stars of the heavens and their constellations
>> will not give their light;
> the sun will be dark at its rising,
>> and the moon will not shed its light. (Is. 13:9-10)

Even more than the light of the divine being removed from those who reject his will, divine light begins to give heat and is embodied in fires that emanate an intolerable burning, wreaking destruction on a wanton world.

> The light of Israel will become a fire,
>> and his Holy One a flame,
> and it will burn and devour
>> his thorns and briers in one day. (Is. 10:17)

The light of the divine, which is an unspeakably great joy and incomparable beauty to the faithful, is to the unfaithful as light is to darkness. The reality of the light chases away the darkness, showing it as the sham it is, revealing that the life and existence of the darkness could only abide so long as divine light gave no reality to the void.

> Let the restless and wicked depart and flee from you. You see them and pierce their shadowy existence: even with them everything is beautiful, though they are vile. What injury have they done you? Or in what respect have they diminished the honour of your rule, which from the heavens down to the uttermost limits remains just and intact? Where have those who fled from your face gone? Where can they get beyond the reach of your discovery? But they have fled that they should not see you, though you see them, and so in their blindness they stumble over you; for you do not desert

anything you have made. The unjust stumble over you and are justly chastised. Endeavoring to withdraw themselves from your gentleness, they stumble on your equity and fall into your anger. They evidently do not know that you are everywhere. No space circumscribes you. (Augustine, Confessions 72)

In contrast with the utter reality of the divine, anything that disconnects itself from God's life-giving influence loses a portion of its existence. It becomes shadowy and unreal, constantly confronted in this world with the physicality of the creative energies that permeate the universe. There is no matter or space, no form or content, without those creative energies. Separation from the divine is death, a non-existence that man affects through his perverted will, a will contrary to that of the divine.

However, if a woman or man might leave the profanities of physical sin, she or he will find that the divine is generous and ready to imbue his beloved with a spiritual reality and physicality that will never lose its depth and weight and density of existence. All major religions agree on this simple, yet profound concept.

> But if a wicked person turns away from all his sins that he has committed and keeps all my statutes and does what is just and right, he shall surely live; he shall not die. None of the transgressions that he has committed shall be remembered against him; for the righteousness that he has done he shall live. Have I any pleasure in the death of the wicked, declares the Lord GOD, and not rather that he should turn from his way and live?... For I have no pleasure in the death of anyone, declares the Lord GOD; so turn, and live. (Ez. 18:21-23,32)

From Buddhism: "Having left the dark way, the skilled person should cultivate the bright... such bright ones, impulses destroyed, are, in this very world, unbound" (Dhammapada 6:87,89). Those who receive glorification find themselves in the boundless realities of eternity, experiencing all reality in one incessant present of utter being and light. The physical impurities of our earthly life are shaken off and left behind.

> If we let Him – for we can prevent Him, if we choose – He will make the feeblest and filthiest of us into a god or goddess, a dazzling, radiant, immortal creature, pulsating all through with such energy

and joy and wisdom and love as we cannot now imagine, a bright stainless mirror which reflects back to God perfectly (though, of course, on a smaller scale) His own boundless power and delight and goodness. (Lewis, Mere Christianity 109)

His divine power has granted to us all things that pertain to life and godliness, through the knowledge of him who called us to his own glory and excellence, by which he has granted to us his precious and very great promises, so that through them you may become partakers of the divine nature, having escaped from the corruption that is in the world because of sinful desire. (2 Peter 1:3-4)

When a man truly and perfectly says with Jesus, and as Jesus said it, 'Thy will be done,' he closes the everlasting life-circle; the life of the Father and the Son flows through him; he is a part of the divine organism. Then is the prayer of the Lord in him fulfilled: 'I in them and them in me, that they may be made perfect in one.' (MacDonald 212)

Man can only access the glory of the divine by giving up control of the divine spark within him, grafting himself into the greater reality of our universe and allowing the life and light of the divine creator to infuse him with the unspeakable truths of an incomprehensible and beautiful reality, laden with enigma and profundity, the likes of which no earthbound being can appreciate in our present physical, imperfect form. We must rely on our will, directing it toward the divine will, allowing it to be shifted and formed by grace and mercy, changing man into a being of immense beauty and blinding radiance. The will is the vital thing.

And taking your life as a whole, with all your innumerable choices, all your life long you are slowly turning this central thing either into a heavenly creature or into a hellish creature: either into a creature that is in harmony with God, and with other creatures, and with itself, or else into one that is in a state of war and hatred with God, and with its fellow-creatures, and with itself. (Lewis, Mere Christianity 55)

Even more than our own eternal state, it may affect some people more acutely to consider our will and our life's influence on our fellow man and his eternal state. The more charitable of us will feel deeply for the destiny of our mother, brother, friend, daughter, nephew, or any other dear relation.

> It may be possible for each to think too much of his own potential glory hereafter; it is hardly possible for him to think too often or too deeply about that of his neighbor. The load, or weight, or burden of my neighbor's glory should be laid on my back, a load so heavy that only humility can carry it, and the backs of the proud will be broken. It is a serious thing to live in a society of possible gods and goddesses, to remember that the dullest and most uninteresting person you can talk to may one day be a creature which, if you saw it now, you would be strongly tempted to worship, or else a horror and a corruption such as you now meet, if at all, only in nightmare. All day long we are, in some degree, helping each other to one or other of these destinations. It is in the light of these overwhelming possibilities, it is with the awe and circumspection proper to them, that we should conduct all our dealings with one another, all friendships, all loves, all play, all politics. There are no *ordinary* people. You have never talked to a mere mortal. (Lewis, The Weight of Glory 45-6)

The will, in its incessant turnings, is directing itself and affecting other lives to move toward or away from the divine light. We are all living eternal realities now in every single passing moment, defining our own reality and the reality of the entire universe forever. Were it not for divine sustenance and support, we would be crushed by the responsibility of every single choice we make, but as it is, we are able to rest in the divine grace and mercy given so freely to those who seek help. So too, we have been given a portion of the divine, aside from the glory of our own will, to guide and direct us toward glorification. Christian theology teaches that the third portion of the divine enigma of the trinity has been given to us as the divine will written on our hearts. No longer must we attend to every dot and tittle in the law, but God himself leads us in the way that we should go, so long as we have the humility to allow him to carry our burden.

> Behold, the days are coming, declares the LORD, when I will make a new covenant with the house of Israel and the house of Judah... I will put my law within them, and I will write it on their hearts. And I will be their God, and they shall be my people. And no longer

shall each one teach his neighbor and each his brother, saying,
'Know the LORD,' for they shall all know me, from the least of them
to the greatest, declares the LORD. For I will forgive their iniquity,
and I will remember their sin no more.
Thus say the LORD,
who gives the sun for light by day
and the fixed order of the moon and the stars for light by night.
(Jer. 31:31,33-35)

The Holy Spirit of God, written on our hearts and directing the reality
of all existence, is of supreme importance for the human will. It gives life and
guides and directs. Through the mysterious workings of the divine will, the
uncertainty of human will is directed down a certain path of personal
glorification and positive corporate influence. By accepting the supremacy of
the divine will, man is able to rid himself of the contingencies of our physical
world and accept the glories of a world of light, of eternal present, of perfect
communion with the divine source of all life. Thus was Jesus able to pray, as
he taught his disciples, with all earnestness and joy, "Your kingdom come,
your will be done, on earth as it is in heaven" (Matt. 6:10).

DARKNESS & VOID

The universe is such a wonderfully rich and complex place that the discovery of the final theory... would not spell the end of science... The ultimate theory would provide an unshakable pillar of coherence forever assuring us that the universe is comprehensible place. (Greene 17)

No theory of life, existence, energy, and physicality proposed outside of the practice of science will ever satisfactorily meet the formalism and testable verifiability of a truly scientific hypothesis. Naturally, much of what would be asserted would be unscientific, untestable, irreplicable, and perhaps physically unreal. Theories of *all* reality will naturally go beyond the boundaries of science, for understanding what underlies physical realities requires information inaccessible to the physical world.

All of the correlations between philosophy, theology, and physics discussed so far in this current text are only correlations. There is little hope to unify these things in a grand collection of laws. The concept of cause and effect, the expectation to find direct relationship, is difficult within a single field of academia and may not even be theoretically possible between fields. Philosophers are playing a losing game if they expect to assert undeniable universal truths. At best, they might find relationship between particular truths.

All this book proposes is relationship. But in those relationships, though we cannot be sure of the direct connection of ideas, we can gain a certain confidence that what we understand and believe will not be falsified by philosophical logic, scientific fact, or religious conviction. As in science, we have a responsibility to accept a theory if we cannot disprove it, it offers compelling explanations for a broad range of phenomena, it is self-consistent, and it is consistent with the balance of our understanding. As in science, in order to accept such a theory, we might have to go through a difficult paradigm shift, ridding ourselves of old misinterpretations of the facts. The transition might not be comfortable, but we cannot disregard our experience, logic, and faith.

If Bohr's position is that such questions can never be satisfactorily formulated, let alone answered, then he seems to be saying that inquiring into the birth of the cosmos is beyond the scope of science. This, to physicists today, simply won't do. (Lindley 217)

As discussed previously, in the later years of his career, Niels Bohr lost much of the conviction that he brought with him into science. The massive shifts in atomic, quantum, and relativistic theory caused him to question what knowledge and what certainties might be accessible through science. It began to appear to this physicist that there are very set, defined limits of what might be known about our world, and they were disappointingly narrow.

Most scientists, past and present, reject this concept, if not in theory, at least in practice. Science goes on, new discoveries happen every day, and we are constantly gaining more knowledge. And although the lower limits of our universe cannot yet be observed directly, through mathematics and indirect observation, the legitimacy of string theory is being tested and is being reformulated into an ever more likely set of principles. Whether or not string theory will ever be practically verifiable, we cannot yet say and have ample reason to doubt. Physicists continue to probe its mathematical possibilities in an attempt to rectify the enigma and impossibilities of quantum physics, special relativity, and our understanding of light. Until some significant progress is made, we all must deal with the fact that we just do not know what we would like to know. The universe is a mysterious place at its core, perhaps ultimately unknowable.

String theory is attempting to rectify the incongruency of relativity and quantum theory by imposing a lower limit on energy, space, and time. This limit rids the physicist of the nonsensical mathematical consequences that emerge when we try to unify quantum theory with relativity. Even more, the relation of these units predicts the measured reality of other natural constants, such as the speed of light in a vacuum. These lower limits are referred to as the Planck units: units of time, space, mass, etc.

The incompatibility of general relativity and quantum mechanics…
is avoided in a universe that has a lower limit on the distances that can be accessed, or even said to exist, in the conventional sense. Such is the universe described by string theory. (Greene 164-5)

The reason for these limits, expressed in our understanding of string theory, is the very fabric of our universe. The string theorist believes that what composes all elementary particles; photons, electrons, gravitons, etc.; are one-dimensional strings which vibrate with incessant energy and existence. The ways in which these strings vibrate determine the elementary particle they express and create.

> String theory proclaims, for instance, that the observed particle properties... are a reflection of the various ways in which a string can vibrate. Just as the strings on a violin or on a piano have resonant frequencies at which they prefer to vibrate... the same holds true for the loops of string theory. (Greene 15)

More, the strings themselves are limited by Planck units. Nothing below the Planck length can possibly exist, so these strings would be the truly elementary physicality of our universe. According to string theory, we cannot say that space and time and any measurement is absolutely continuous, for the continuity stops at the Planck unit. Nothing below those limits are accessible. Perhaps nothing below those limits is even real.

Of great importance, one consequence of string theory is that there is no way to conceive of void or darkness or absence. The vibrating strings themselves create the reality of space and time, without which there would be nothing, that is, not empty space, but no space at all; not an absence of time, but no reality of time to begin with. The strings give reality to reality and anything else *is* not. This characteristic of string theory helps the scientist, and all those with the ability to understand, to conceptualize the most difficult propositions of philosophy and theology: death, hell, and non-existence.

THE THEORY OF THE VOID

And thus, as long as I think only about God and focus completely on him, I find no cause of error or falsehood in myself. But as soon as I turn back to myself, however, I find that I am subject to innumerable errors. When I look for a cause of these errors, I find that I have not only a real and positive idea of God or of a supremely perfect being but I also have, if I may so describe it, a certain negative idea of nothingness or of what is removed as far as possible from every perfection. (Descartes 45)

'A negative idea of nothingness' is a human concept that we use to try to understand what existence without form or substance might be, which of course, could not be, in our commonsense understanding. When we talk about nothingness, we are not being scientific at all. What we describe is not a physical phenomenon, but a mental construct.

> Privations – rest, for example – are also something intellectual. "Rest" is a term for the absence of motion in what may be conceived to be in motion, and absence is not something realized in concrete things but is something intelligible in the mind. (Suhrawardi 49)

Down to the level of quantum particles and the strings that might form them, we find that there is no reality of 'rest'. The strings themselves are vibrating constantly, an integral part of their existence, so that we cannot properly say that anything composed of these strings (i.e. everything) is ever at rest. Privations have no physical reality.

Even theological privations are far more slippery than many have ever imagined. We think we understand spiritual darkness, the displeasure of God, and hell itself, but when we try to put flesh on these concepts, our best efforts are foiled by the shadowy unreality of privations. Job said as much when he considered the philosophies of those around him who did not believe in God.

> They make night into day:
>> 'The light,' they say, 'is near to darkness.'
> If I hope for Sheol as my house,
>> if I make my bed in darkness,
> if I say to the pit, 'You are my father,'
>> and to the worm, 'My mother,' or ' My sister,'
> where then is my hope?
>> Who will see my hope? (Job 17:12-15)

In these statements, Job was trying to conceptualize how his neighbors could believe that there was no ultimate divinity that energized and vitalized creation. By ridding oneself of that divine source, a man is left with no reality for want of a source of reality and no hope for want of a source of hope. Job lamented the evil that he saw around himself, and rightly understood the perversion of will that he witnessed as an absence of the life-giving influence of the divine. William Paul Young understands evil in the same terms, in an imagined statement from God:

> *Evil* is a word we use to describe the absence of good, just as we use the word *darkness* to describe the absence of light or *death* to describe the absence of life. Both evil and darkness can be understood only in relation to light and good; they do not have any actual existence. I am light and I am good. I am love and there is no darkness in me. Light and Good actually exist. So, removing yourself from me will plunge you into darkness. Declaring independence will result in evil because apart from me, you can draw only upon yourself. That is death because you have separated yourself from me: Life. (Young 138)

The light and life of the divine is a self-existent thing, sharing its goodness and energies with those who have the ability and responsibility to accept it, though we might ultimately reject that life. Sin, as theologically understood, is nothing but the willful rejection of the divine light, the absence of goodness.

> The movement of the will away from you, who are, is movement towards that which has less being. A movement of this nature is a fault and a sin, and no one's sin harms you or disturbs the order of your rule. (Augustine, Confessions 251)

The goodness of the divine cannot be affected by those who would choose to turn from it. God's existence remains steadfastly self-existent; however, the life of he who would turn from life itself has little hope for continuation. There are natural and entirely unavoidable consequences for such a turning.

> Bold and willful... these, like irrational animals, creatures of instinct, born to be caught and destroyed, blaspheming about matters of which they are ignorant, will also be destroyed in their destruction, suffering wrong as the wage for their wrongdoing. (2 Peter 2:10,12-13)

One who abandons life cannot be brought to life, and any existence that he can be said to experience will be a thing with thin reality. Death and hell, though they are real consequences of particular actions, are unreal in themselves, in a troublingly real way. Death and hell are privations of life and good, things without form or content in themselves, but acutely experienced by someone who has been removed from the physicality of the divine.

> Hell is a state of mind – ye never said a truer word. And every state of mind, left to itself, every shutting up of the creature within the dungeon of its own mind – is, in the end, Hell. But Heaven is not a state of mind. Heaven is reality itself. All that is fully real is Heavenly. (Lewis, The Great Divorce 338)

> Life is the only reality; what men call death is but a shadow – a word for that which cannot be – a negation, owing the very idea of itself to that which it would deny. But for life there could be no death. If God were not, there would not even be nothing. Not even nothingness preceded life. Nothingness owes its very idea to existence. (MacDonald 151)

The stark reality of existence is unquestionable and the lifelessness of evil is a natural departure of the energies of the divine. Sin, death, and hell walk hand-in-hand away from the ultimate source of *be*, and so they are not, leading those who reject the divine will toward a tortuous reality of shadowy lifelessness. Even the cause of evil seems to lack the tangible reality that we expect from actions that cause certain consequences in turn. We know what evil is, a departure from the divine will, but from whence it comes and why it *is* is impossible to discern. "To try to discover the causes of such defection – deficient, not efficient causes – is like trying to see darkness or to hear silence" (Augustine, City of God 480). Sin lacks the reality of life, though it does not lack reality. The sin is real, its existence is 'shadowy'. The Qur'an asserts as much:

> But the deeds of disbelievers are like a mirage in a desert... Or like shadows in a deep sea covered by waves upon waves, with clouds above – layer upon layer of darkness – if he holds out his hand, he is scarcely able to see it. The one to whom God gives no light has no light at all. (Qur'an 24:39-40)

The existence of evil is a paradox of impossible difficulty. Man cannot understand what existence without life might be. We do not understand how a reality without form, substance, or creative energies could be. Yet, we have all experienced this absence of goodness. We feel the absence, though we cannot put our finger on it and experiment on it and consider it in itself. Evil's existence affects its death. Evil is the tendency toward death, away from life.

> Herein lies the chief irony of sin: it wants distance from God, it desires autonomy from the Creator, but in this distance is found its own destruction. When cancer destroys its host, it simultaneously destroys itself. Evil is a cancer out to destroy God's good world, and it doesn't care if it goes down with the ship. Sin seeks to drag creation back into the nihilistic void from which it came. If evil had its way, it would no longer be evil; it would cease to exist... As evil seeks distance from God, the Light of the world, it naturally engrosses itself ever further in the darkness of its own autonomy. (Butler 58)

Sin, man's distance from the divine, is a lifelessness and a void that is incomprehensible in its unreality, though it is the real consequence of a perverted will. The divine is the only reality, the only life accessible in our universe. So, if sinful man is to ask what his own existence is, he might find himself in a difficult space. We know sin is death and we know that the divine is life, but man seems to live in this enigmatic no man's land of the universe, split between the perfection of the creator and the non-existence of life apart from him.

> But who am I, and what is my people, that we should be able thus to offer willingly? For all things come from you, and of your own have we given you. For we are strangers before you and sojourners, as all our fathers were. Our days on earth are like a shadow, and there is no abiding. (1 Chron. 29:14-15)

King David of Israel is recorded to have said this before his death, and his own confrontation of mortality illuminates what we should all consider. Somehow the human will is a self-existent thing, owing its creation to the divine but continuing into perpetuity as an independent source of activity. All life is derived from the divine, but so long as the present activates human will, human life has a certain continuity that defies the natural reliance of life on the divine. We are imbued with the energy and goodness of God, but also given the right and responsibility to continue living in him. If we do not, we are removed from the source of life, though we are told that our soul goes on in some horrendous existence of total privation and death. The way that man is created, sustained, and yet somehow independent of God is similar to that of the angels, if we are to believe a Judeo-Christian understanding of Satan and his demons.

> For when God said, 'Let there be light', and light was created, then, if we are right in interpreting this as including the creation of the angels, they immediately become partakers of the eternal light, which is the unchanging Wisdom of God, the agent of God's whole creation; and this Wisdom we call the only begotten Son of God... This is the true light, which illuminates every man as he comes into this world; and this light illuminates every pure angel, so that he is not light in himself, but in God. If an angel turns away from God he becomes impure: and such are all those who are called 'impure spirits'. They are no longer light in the Lord; they have become in themselves darkness, deprived of participation in the eternal light. For evil is not a positive substance: the loss of good has been given the name of 'evil'. (Augustine, City of God 439-40)

The creative and life-sustaining force of the divine and its goodness expressed in our universe meets its unequal and unreal antithesis in sin, evil, death, and Satan. There is no way for man to understand this reality except in its relation to the divine good.

> [Sin, death, and hell] constitute an "anti-creation" force, not as substantive things in themselves so much as parasites that prey upon the good creation God has made in an attempt to devour it and destroy it, to drag creation back into the nothingness, the darkness, the void from which it came. (Butler 11)

The unreality of sin and hell is emphasized in an important observation made by pastor Joshua Ryan Butler. In his book, *The Skeletons in God's Closet*, Butler asserts that a concordance search of the Bible will show that the terms *heaven* and *hell* appear in no verses together. That's right. The dichotomy that all of us so readily assume defines the character of our universe, the Yin and Yang, good and evil, heaven and hell are not two sides of the same coin.

What we had always assumed to be the two sides of reality are hardly even related. A concordance search of the Bible will show, conversely, heaven and earth appearing together nearly 200 times. When the divinity of our universe created all reality, he created the heaven of his own existence and the earthliness of ours. Hell and death were not a part of this divine duality.

Hell enters into the world as a privation, as a parasite of the goodness of the divine, antithetical to divine plans. Sin is an absence, lacking the physicality of the light and energy of goodness, the will of God and the obedience of man. Sin is "the world of darkness, which is but a shadow of the world of light" (Suhrawardi 107). Light and dark are not corollaries, but opposites.

> But the path of the righteous is like the light of dawn,
>> which shines brighter and brighter until full day.
> The way of the wicked is like deep darkness;
>> they do not know over what they stumble. (Prov. 4:18-19)

Still, what the existence of non-existence is remains a great mystery in philosophy and theology. Science rids itself of such silly notions by reminding us that science only ever deals with that which might be observed, and the privation of existence cannot have any physical reality to be observed. Still, philosophically and theologically, we know that there must be some sort of reality to privations, though there is no way for us to comprehend such a concept.

> Now since the soul, being created immortal, cannot be deprived of every kind of life, the supreme death of the soul is alienation from the life of God in an eternity of punishment. (Augustine, City of God 253)

> A corruptible being may be destroyed even though its emanating cause remains because it depends upon other causes that have ceased to be. (Suhrawardi 123)

Hell will be the reward of those who return to their Lord as evildoers: there they will stay, neither living nor dying. (Qur'an 20:74)

The existence of sin and hell are truly incomprehensible when we consider the logical corollaries of a supreme and sovereign divinity; however, we somehow understand the glory and reality of the human will and its paradoxical effects on the creation of a good creator. Perhaps the best way to conceptualize the dichotomy is to appreciate the Biblical understanding of the redeemed and sanctified city of Jerusalem and its role in the new order of things, after sin and death are removed from God's good reality. The city and its internal goodness are contrasted against the darkness that is barred from entry through the heavenly gates.

Inside are the lights and laughter that mark the communion of grace. Outside are the darkness and tears that mark the vanquished self-reliance of sin. (Butler 170)

It is the vast outside; the ghastly dark beyond the gates of the city of which God is the light... The man wakes from the final struggle of death, in absolute loneliness – such a loneliness as in the most miserable moment of deserted childhood he never knew. Not a hint, not a shadow of anything outside his consciousness reaches him. All is dark, dark and dumb; no motion..." (MacDonald 135)

If any man is sure of his final 'resting' place, let him consider again the obedience that is required of him and his will. The divine will commands that the human will be aligned with it. He who is self-righteous and sure of his own goodness is encouraged to revisit his motivations and desires and actions, ensuring that he lacks nothing of the divine essence.

And of course, upon reflection, any created thing will recognize that he lacks at least a portion of the essence of the creator. The creation cannot be wholly like its creator, but the creation can hope that its creator might have the mercy and grace to accept its flaws and to revitalize the reality of the creation into a more perfect existence. All creation seeks a more *real* reality. We crave the light. "Let us not pass away from light to exile" (Rig 2:28.7). Those are our options: not perfect light or less light, but perfect light or complete and utter removal from any light. And the self-righteous might be surprised that they find themselves removed from the perfection of reality and existence.

> I tell you, many will come from east and west and recline at table with Abraham, Isaac, and Jacob in the kingdom of heaven, while the sons of the kingdom will be thrown into the outer darkness. (Matt. 8:11-12)

Let us not be sons and daughters of the kingdom, heirs of an impotent lineage, but let us inherit light, eternally present and active and contributing to the greater goodness of a perfectly good reality. We rely on the grace of such goodness, knowing that the human element of our will must continually tear us away from the perfection of complete existence, always affecting our own tendency toward death.

"Grace doesn't depend on suffering to exist, but where there is suffering you will find grace in many facets and colors" (Young 187-8). Grace is a reality in itself, rescuing the privations of our own self-reliance from the non-existence of humanity, death, and hell. Grace imbues the profane with the light and energy and existence and reality of the divine. Only after the redemption, sanctification, and glorification of our mortal souls will the ultimate and enduring reality of our eternal existence be affected: "Then the righteous will shine like the sun in the kingdom of their Father" (Matt. 13:43). Only in the sovereign grace of a good god can any mortal hope to attain the abiding light of life.

CONCLUSION

What have I said that has not already been said? The information presented in this book has been studied in depth in many other works written by far more intelligent men, true experts of the individual fields of science, philosophy, and theology. If the quotes on every page have not driven the fact home, let me assure you, not much of what was discussed here is novel. It is possible that those who would read the same sources would come to the same conclusions. What I have done is little more than to relate ideas that are seldomly considered together, which certainly appear to be intimately related.

And even in the amalgamation of all these ideas, there are necessarily truths which I have overlooked and those on which I placed little importance, which are integral to the reality of the world in which we live. In the words of Wittgenstein, "My thoughts probably move in a far narrower circle than I suspect" (Wittgenstein, Culture and Value 63e).

However, if the reader is displeased with the result of this text, he or she does not share the feelings of the author. I expressed nearly all that I wished (of which I was able). I learned much about God's will and my own will, its importance and responsibility, in the research required to write this book. I learned that the sovereignty of God is not at all jeopardized by the defection of my own will, and I learned about the gravity of the glory that I bear.

Of utmost importance, I gained greater appreciation for the fact that God can say of me right now and always, "Well done, good and faithful servant." The eternal divinity of our universe does not see us as we are at any single moment, but as we are eternally. God sees me in my redeemed state. God sees me as his perfect creation, not as a sin-riddled mortal man defined by my past. The eternally present reality of God assures that I am only ever perceived in his eternal and perfect light, the state affected by the grace and mercy bled out upon the cross of Jesus.

I am defined by a single thing. If someone were to ask me, "How do you identify?", I could respond with my political bent, my sexuality, or my

251

current profession. Instead, if I were to answer with eternity in mind, as I always should, I would say, "God is good, and he loves me." My true identity can only ever be expressed in relation to the divine character. No other identity would be as real. We all ought to pray continually that God show us his divine will and that we might have the propensity to abide by it.

> All that is needed to set the world right enough for me... is, that I care for God as he cares for me; that my will and desires keep time and harmony with his music; that I have no thought that springs from myself apart from him; that my individuality have the freedom that belongs to it as born of his individuality, and be in no slavery to my body, or my ancestry, or my prejudices, or any impulse whatever from region unknown; that I may be free by obedience to the law of my being, the live and live-making will by which life is life, and my life is myself. What springs from myself and not from God, is evil; it is a perversion of something of God's. Whatever is not of faith is sin; it is a stream cut off – a stream that cuts itself off from its source, and thinks to run on without it. But light is my inheritance through him whose life is the light of men, to wake in them the life of their father in heaven. Loved be the Lord who in himself generated that life which is the light of men! (MacDonald 301)

If any reader is still determined to live a good life and thereby earn the rewards of the eternal Father, his focus is on the ends instead of the means. If it has not been abundantly affirmed at this point, let me reemphasize that the ends and the means are one and the same from the eternal perspective of the inexorable present. We are living eternity NOW! We are not waiting to be whisked up into a perfect heaven. We are living in and contributing to the earthly kingdom of God which is being and will be completely redeemed. The present is not concerned with ends, but with reality, and the man who lives obediently in the present abides by the eternal will of that reality. He who truly seeks the divine lives into it.

> If you want joy, power, peace, eternal life, you must get close to, or even into, the thing that has them. They are not a sort of prize which God could, if He chose, just hand out to anyone. They are a great fountain of energy and beauty spurting up at the very centre of reality. (Lewis, Mere Christianity 96)

CONCLUSION

The astute reader who accepts the reality of the truths expounded in this text, if she does not already know, will ask, of course, "What is the will of God, that I might abide by it?"

There are a few abundantly clear portions of scripture that clarify what the will of God is for man, and these will serve as a good starting point for anyone who wants to abide by the divine will:

> For this is the will of my Father, that everyone who looks on the Son and believes in him should have eternal life, and I will raise him up on the last day. (John 6:40)

> For this is the will of God, your sanctification. (1 Thess. 4:3)

> For this is the will of God, that by doing good you should put to silence the ignorance of foolish people. (1 Peter 2:15)

The will of God is that you believe in him and his redemptive plan. The will of God is to make you into a being of glorious and perfect light. The will of God is that your goodness, not necessarily your knowledge, will quiet the ignorance of those who do not yet understand.

> Be at peace among yourselves. And we urge you, brothers, admonish the idle, encourage the fainthearted, help the weak, be patient with them all. See that no one repays anyone evil for evil, but always seek to do good to one another and to everyone. Rejoice always, pray without ceasing, give thanks in all circumstances; for this is the will of God in Christ Jesus for you. (1 Thess. 5:13-18)

Of course, as we should all assume at this point, the will of God will not be communicated to the world through any empty words, but through the eternal significance of our present goodness. Though they are obviously useful, earthly knowledge and wisdom are of lesser importance, of little importance, compared to earthly obedience. The reading of this book will have little effect on your eternity if you do not act. The eternal requires action. Action is key.

For those who have been affected by the words and reasoning of this text, let them wrestle with the realities encountered. There is no time like the present. There is no time *but* the present. I ask no one to accept the message of Jesus in blind faith or against his or her own reason. However, if your reason is challenging old assumptions, please, attend to that which is on your mind and heart. It is possible that you are considering issues of eternal significance, and if so, you cannot tolerate a single moment of unnecessary doubt. Latch onto the truths readily available, and work to develop your own understanding of a greater reality. *Be*, lest you *will* not. Reflect to the world every bit of the divine light that has been generously bestowed upon you.

Finally, in the words of the remarkable Saint Augustine:

And now, as I think, I have discharged my debt, with the completion, by God's help, of this... work. It may be too much for some, too little for others. Of both these groups I ask forgiveness. But of those for whom it is enough I make this request: that they do not thank me, but join with me in rendering thanks to God. Amen. Amen. (Augustine, City of God 1091)

CONCLUSION

REFERENCES

Arcand, Kimberly and Megan Watzke. *Light: The Visible Spectrum and Beyond*. Hachette Book Group, 2015.

Aristotle. *The Nicomachean Ethics*. Trans. David Ross. Oxford University Press, 2009.

Augustine. *City of God*. Trans. Henry Bettenson. Penguin Group, 2003.

—. *Confessions*. Trans. Henry Chadwick. Oxford University Press, 2008.

The Bhagavad-Gita. Trans. Barbara Stoler Miller. Random House, 2004.

The Bible. English Standard Version. Crossway Bibles, 2008.

Butler, Joshua Ryan. *The Skeletons in God's Closet*. Thomas Nelson, 2014.

Calvino, Italo. *If on a Winter's Night a Traveler*. Trans. William Weaver. Harcourt, 1981.

Descartes, Rene. *Meditations and Other Metaphysical Writings*. Trans. Desmond M. Clarke. Penguin Group, 2000.

The Dhammapada: Verses on the Way. Trans. Glenn Wallis. Random House, 2007.

Greene, Brian. *The Elegant Universe*. W.M. Norton and Co., 2003.

Hecht, Eugene. *Optics*. 2nd ed. Addison-Wesley, 1987.

Kuhn, Thomas S. *The Structure of Scientific Revolutions*. 4th ed. University of Chicago Press, 2012.

Lewis, C.S. "Is Theology Poetry." *The Weight of Glory and Other Addresses*. Harper Collins, 1980.

—. "Mere Christianity." *The Complete C.S. Lewis Signature Classics*. Harper Collins, 2002.

—. "The Great Divorce." *The Complete C.S. Lewis Signature Classics*. Harper Collins, 2002.

—. "The Weight of Glory." *The Weight of Glory and Other Addresses*. Harper Collins, 1980.

—. *Till We Have Faces*. Harcourt, 2012.

Lindley, David. *Uncertainty: Einstein, Heisenberg, Bohr, and the Struggle for the Soul of Science*. Doubleday, 2007.

MacDonald, George. *Unspoken Sermons*. Classic Reprint Series, CreateSpace, 2016.

March, Robert H. *Physics for Poets*. McGraw-Hill, 1970.

Pedrotti, Frank L. and Leno S. Pedrotti. *Introduction to Optics*. 2nd ed. Prentice Hall, 1993.

Pelton, Miles. *First Law of Physics: Let There Be Light*. Xlibris, 2013.

REFERENCES

Pirsig, Robert M. *Zen and the Art of Motorcycle Maintenance*. HarperTorch, 2006.

Plato. *The Republic*. Trans. Desmond Lee. Penguin Group, 2003.

The Qur'an. Trans. M.A.S. Abdel Haleem. Oxford University Press, 2010.

Rosenblum, Bruce and Fred Kuttner. *Quantum Enigma: Physics Encounters Consciousness*. Oxford University Press, 2011.

Russell, Bertrand. *A History of Western Philosophy*. Simon and Schuster, 1945.

Sagan, Carl. *Pale Blue Dot*. Random House, 1994.

Spinoza, Benedict. *The Chief Works of Benedict de Spinoza*. Trans. R.H.M. Elwes. Kshetra Books, 2016.

Steinbeck, John. *East of Eden*. Penguin Group, 2002.

Suhrawardi. *The Philosophy of Illumination: A New Critical Edition of the Text of Hikmat al-ishraq*. Trans. John Walbridge and Hossein Ziai. Brigham Young University Press, 1999.

Taylor, Jill Bolte. *My Stroke of Insight*. Penguin Group, 2006.

Thoreau, Henry David. *Walden*. Barnes and Noble, 2004.

The Vedas: The Samhitas of the Rig, Yajur (White and Black), Sama, and Atharva Vedas. Trans. Ralph T.H. Griffith and Arthur Berriedale Keith. Kshetra Books, 2017.

Wittgenstein, Ludwig. *Culture and Value*. Ed. G.H. Von Wright. Trans. Peter Winch. University of Chicago Press, 1984.

—. *Philosophical Investigations*. Trans. G.E.M. Anscombe. 3rd ed. Basil Blackwell & Mott, 1958.

Young, William Paul. *The Shack*. Windblown Media, 2007.